W0055450

Algebraic Systems of Equations and
Computational Complexity Theory

Mathematics and Its Application (China Series)

Managing Editor:

M. HAZEWINKEL

Centre for Mathematics and Computer Science,
Amsterdam, The Netherlands

Algebraic Systems of Equations and Computational Complexity Theory

Wang Zeke
Zhongshan University
Guangzhou, The People's Republic of China

Xu Senlin
University of Science and Technology of China
Hefei, The People's Republic of China

Gao Tangan
Zhongshan University
Guangzhou, The People's Republic of China

 Science Press
Beijing

 SPRINGER-SCIENCE+BUSINESS MEDIA, B.V.

Library of Congress Cataloging in Publication Data

Wang, Tse-k'o, 1942–

Algebraic systems and computational complexity theory / by Z, Wang and S. Xu and T. Gao.

p. cm. — (Mathematics and its applications : v. 269)

Includes bibliographical references and index.

ISBN 978-94-010-4342-7   ISBN 978-94-011-0796-9 (eBook)
DOI 10.1007/978-94-011-0796-9

1. Homotopy theory. 2. Equations–Numerical solutions. 3. Computational complexity. I. Hsü, Sen-lin. II. Kao, Tang-an, 1964– III. Title. IV. Series: Mathematics and its applications (Kluwer Academic Publishers): v. 269.
QA612.7.W35 1993
514'.24–dc20   93-31541

ISBN 978-94-010-4342-7

All Rights Reserved
©1994 by Springer Science+Business Media Dordrecht
Originally published by Kluwer Academic Publishers in 1994
Softcover reprint of the hardcover 1st edition 1994
No part of the material protected by this copyright notice may be reproduced or utilized in any form or by any means, electronic or mechanical, including photocopying, recording, or by any information storage and retrieval system, without written permission from the copyright owners.

SERIES EDITOR'S PREFACE

'Et moi, ..., si j'avait su comment en revenir, je n'y serais point allé.'

Jules Verne

The series is divergent; therefore we may be able to do something with it.

O. Heaviside

One service methematics has rendered the human race. It has put common sense back where it belongs, on the topmost shelf next to the dusty canister labelled 'discarded nonsecse'.

Eric T. Bell

Mathematics is a tool for thought. A highly necessary tool in a world where both feedback and nonlinearities abound. Similarly, all kinds of parts of mathematics serve as tools for other parts and for other sciences.

Applying a simple rewriting rule to the quote on the right above one finds such statements as: 'One service topology has rendered mathematical physics ...'; 'One service logic has rendered computer science ...'; 'One service category theory has rendered mathematics ...'. All arguable true. And all statements obtainable this way form part of the raison d'être of this series.

This series, *Mathematics and Its Applications*, ttarted in 1977. Now that over one hundred volumes have appeared it seems opportune to reexamine its scope. At the time I wrote

"Growing specialization and diversification have brought a host of monographs and textbooks on increasingly specialized topics. However, the 'tree' of knowledge of mathematics and related fields does not grow only by putting forth new branches. It also happens, quite often in fact, that branches which were thought to be completely disparate are suddenly seen to be related. Further, the kind and level of sophistication of mathematics applied in various sciences has changed drastically in recent years: measure theory is used (nontrivially) in regional and theoretical economics; algebraic geometry interacts with physics; the Minkowsky lemma, coding theory and the structure of water meet one another in packing and covering theory; quantume fields, crystal defects anf mathematical programming profit from homotopy theory; Lie algebras are relevant to filtering; and prediction and electrial engineering can use Stein spaces. And in addition to this there are such new emerging subdisciplines as "experimental methematics', 'CFD', 'completely integrable systems', 'chaos, synergetics and large-scale order', which are almost impossible to fit into the existing classification schemes. They draw upon widely different sections of mathematics."

By and large, all this this still applies today. It is still true that at first sight mathematics seems rather fragmented and that to find, see, and exploit the deeper underlying interrelations more effort is needed and so are books that can help mathematicians and scientists do so. Accordingly MIA will continue to try to make such book available.

If anything, the description I gave in 1977 is now an understatement. To the examples of interaction areas one should add string theory where Riemann surfaces, algebraic geometry, modular functions, knots, quantum field theory, Kac-Moody algebras, monstrous moonshine (and more) all come together. And to the examples of things which can be usefully applied let me add the topic 'finite geometry'; a combination of words which sounds like it might not even exist, let alone be applicable. And yet it is being applied: to statistics via designs, to radar/sonar detection arrays (via finite projective planes), and to bus connections of VLSI chips (via difference sets).

There seems to be no part of (so-called pure) mathematics that is not in immediate danger of being applied. And, accordingly, the applied mathematician needs to be aware of much more. Besides analysis and numerics, the traditional workhorses, he may need all kinds of combinatorics, algebra, probability, and so on.

In addition, the applied scientist needs to cope increasingly with the nonlinear world and the extra mathematical sophistication that this requires. For that is where the rewards are. Linear models are honest and a bit sad and depressing: proportional efforts and results. It is in the nonlinear world that infinitesimal inputs may result in macroscopic outputs (or vice versa). To appreciate what I sm hinting at; if electronics were linear we would have no fun with transistors and computers; we would have no TV; in fact you would not be reading these lines.

There is also no safety in ignoring such outlandish things as nonstandard analysis, superspace and anticommuting integration, p-adic and ultrametric space. All three have applications in both electrical engineering and physics. Once, complex numbers were equally outlandish, but they frequently proved the shortest path between 'real' results. Similarly, the first two topics named have already provided a number of 'wormhole' paths. There is no telling where all this is leading-fortunately.

Thus the original scope of the series, which for various (sound) reasons now comprises five subseries: white (Japan), yellow (China), red (USSR), blue (Eastern Europe), and green (everything else), still applies. It has been enlarged a bit to include book treating of the tools from one subdiscipline which are used in others. Thus the series still aims at books dealing with:
- a central concept which plays an improtant role in several different mathematical and/or scientific specialization areas;
- New applications of the results and ideas from one area of scientific endeavour into another;
- influences which the results, problems and concepts of one field of enquiry have, and have had, on the development of another.

The present volume, one of the first in the 'Chinese subseries' of MIA, also appropriately enough, one dealing with fundamental issues: interrelations between logic and computer science. The advent of computers has sparked off revived interest in a host of fundamental issues in science and mathematics such as computability, recursiveness, computational complexity and automated theorem proving to which latter topic ths author has made seminal contributions for which he was awarded the ATP prize in 1982.

It is a pleasure to welcome this volume in this series.

The shortest path between two truths in the real domain passes through the complex domain

J. Hadamard

La physique ne nous donne pas seulement ;'occasion de rèsoudr des problèmes ... elle nous fait presentir la solution.

H. Poincaré

Never lend books, for no one ever returns them; the only books I have in my library are books that other folk have lent me.

Anatole France

The function of an expert is not to be more right than other people, but to be wrong for more sophisticated reasons.

David Butler

Bussum, August 1989

Michiel Hazewinkel

Preface

In the past fifteen years, significant progress has been made in the solution of nonlinear systems, particularly in the areas of computing fixed points, solving nonlinear equation systems, and in applying these methods to equilibrium models. This progress has developed along two principal lines : simplicial and continuation methods. Simplicial methods stem from the pioneering work of Scarf on the approximation of fixed points. As shown by Kuhn, the essential idea is the use of a simplicial approximation of the map as is used in the proof of Brouwer's fixed point theorem by Sperner's lemma. Continuation methods originate in the the work of Kellogg, Li, and Yorke which converted the nonconstructive proof by Hirsch of Brouwer's theorem into a constructive algorithm.

The development of these two lines of approach has been parallel in many respects. For example, both have used the idea of a homotopy to make a transition from an easy problem to a difficult problem. On the other hand, the two points of view are contrasted in the behavior of numerical algorithms, where continuation methods normally work with probability near one (excluding "bad" cases) while simplicial methods frequently work without exception.

The book by Wang and Xu presents a self-contained exposition of recent work on simplicial and continuation methods applied to the solution of algebraic equations. For the case of the search for the roots of a single polynomial over the complex field, the simplicial algorithm studied is that proposed by Kuhn. Wang and Xu give a complete and self-contained exposition of this algorithm. This is followed by a discussion of error, cost, and efficiency, areas to which Wang and Xu have made original and interesting contributions. For the same problem approached by continuation methods, the starting point is Smale's recent study of a global Newton method.

Their exposition of Smale's work is notable for its clarity; in addition, they have corrected numerous errors in the only published version available and have filled several gaps in the arguments. This discussion is followed by original research that compares the cost estimates for Kuhn's algorithm and Smale's estimates for Newton's method.

The second half of the book deals with very recent research on systems of algebraic equations. It is notable for its exposition of the various tools from widely different mathematical subjects (such as algebraic, geometry and differential topology) that have been applied to this problem. As in the first part of the book, both continuation and simplicial methods are discussed. The final chapter contains contributions the authors have made to the design of a simplicial homotopy algorithm for the numerical solution of systems of nonlinear algebraic equations.

It is a pleasure to introduce this book to the reader and student. It is certain that, by their careful exposition of this active area of research, Wang and Xu will generate interest and make further progress on these problems possible.

H.W. Kuhn

Contents

Chapter 1

Kuhn's Algorithm for Algebraic Equations

This chapter is devoted to Kuhn's algorithm for algebraic equations and to the proof of its convergence.

The main references are [Kuhn, 1974; 1977].

Contrary to all traditional methods of iteration, Kuhn's algorithm is based on simplicial triangulation of the underlying space, an integer labelling, and a complementary pivoting procedure of computation. If its description is not so simple as, say, Newton method, after certain implementation, its use is, however, tremendously easy. To solve any algebraic equation by using Kuhn's algorithm, the only thing one should do is to input the complete set of its coefficients as well as the accuracy demand into the machine. Then the algorithm will find one by one all solutions with no more care needed. For Kuhn's algorithm, there are no difficult problems like the selection of initial values. It is a method with a very strong guarantee of global convergence.

Finally, for the purpose of only implementing the algorithm, it is enough to know the first two sections.

§1. Triangulation and Labelling

Let $f(x)$ be a monic polynomial of degree n in the complex variable z with complex numbers as its coefficients, that is, $f(z) = z^n + a_1 z^{n-1} + \cdots + a_n$, where n is a positive integer and $a_1, \cdots, a_n$ are complex constants. If a complex number ξ satisfies $f(\xi) = 0$, ξ is called a zero of the polynomial f or a solution of the algebraic equation $f(z) = 0$. Kuhn's algorithm is designed to find all zeros of f.

Denote by $\mathbf{C}$ the plane of complex $z = x + iy$ and by $\mathbf{C}'$ the plane of complex $w = u + iv$. Then $w = f(z)$ defines a polynomial mapping $f : \mathbf{C} \to \mathbf{C}'$.

To describe Kuhn's algorithm in the next section, we now introduce a triangulation of the half-space $\mathbf{C} \times [-1, +\infty)$ and a labelling rule for the vertices of the triangulation.

Let $\mathbf{C}_d$ denote the replica $\mathbf{C} \times \{d\}$ of the plane $\mathbf{C}$ for $d = -1, 0, 1, 2, \cdots$. Then $\mathbf{C}_d \subseteq \mathbf{C} \times [-1, +\infty)$. Given a center $\tilde{z}$ and a grid size h, we define the triangulation $\mathbf{T}$ of $\mathbf{C} \times [-1, \infty)$ as follows.

Triangulation $\mathbf{T}_d(\tilde{z}; h)$ or $\mathbf{T}_d$ of the plane $\mathbf{C}_d$.

The triangulation $\mathbf{T}_{-1}(\tilde{z}; h)$ of $\mathbf{C}_{-1}$ is illustrated in Fig. 1.1. A triangle in $\mathbf{T}_{-1}(\tilde{z}; h)$ is uniquely determined by a pair of integers (r, s) with $r + s$ even and $(a, b) \in \{(1, 0), (0, 1), (-1, 0), (0, -1)\}$. The z-coordinates of its vertices are:

$$\tilde{z} + (r + i\, s)h;$$

$$\tilde{z} + ((r + a) + i\, (s + b))h;$$

$$\tilde{z} + ((r - b) + i\, (s + a))h.$$

The supremum of the diameters of its triangles is called the mesh of the triangulation $\mathbf{T}_{-1}(\tilde{z}; h)$. The mesh of the triangulation $\mathbf{T}_{-1}(\tilde{z}; h)$ is obviously $\sqrt{2}h$.

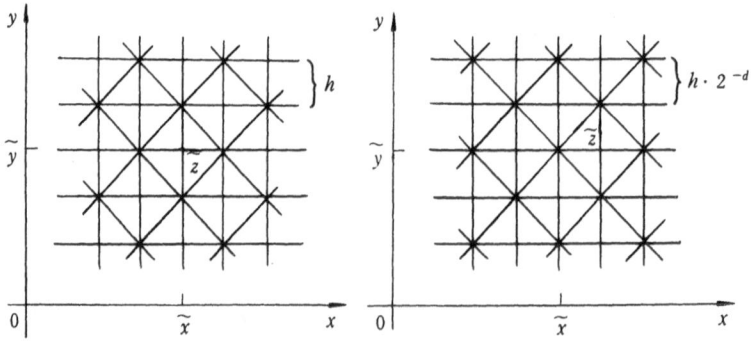

Figure 1.1   Figure 1.2

The triangulation $\mathbf{T}_d(\tilde{z}; h)$ of $\mathbf{C}_d$ is illustrated in Fig. 1.2, where $d = 0, 1, 2, \cdots$. A triangle in $\mathbf{T}_d(\tilde{z}; h)$, where $d \geq 0$, is uniquely specified by an $(a, b) \in \{(1, 0), (0, 1), (-1, 0), (0, -1)\}$ and a pair of integers (r, s) with

$r + s$ odd. The z-coordinates of its vertices are

$$\tilde{z} + (r + i\,s)h2^{-d};$$
$$\tilde{z} + ((r + a) + i\,(s + b))h2^{-d};$$
$$\tilde{z} + ((r - b) + i\,(s + a))h2^{-d}.$$

With a similar definition, the mesh of $\mathbf{T}_d(\tilde{z}; d)$ is $\sqrt{2}h2^{-d}$.

Notice that every triangle in the triangulation $\mathbf{T}_d(\tilde{z}; h)$ is an isosceles right triangle with two right-angle sides paralleling to the x-axis and to the y-axis, respectively.

Remark 1.1 The triangulation $\mathbf{T}_d(\tilde{z}; h)$ can be easily portrayed by four families of parallel lines. But the advantage of the above description is that every vertex in the triangulation is specified by two pairs of integers r, s and a, b.

Triangulation $\mathbf{T}(\tilde{z}; h)$ or $\mathbf{T}$ of the half-plane $\mathbf{C} \times [-1, \infty)$.

By definition of the triangulation $\mathbf{T}_d(\tilde{z}; h)$, for every square in $\mathbf{C}_{-1}$ consisting of exactly two triangles in $\mathbf{T}_{-1}(\tilde{z}; h)$ with a common hypotenuse, there is a unique "above-opposite" square in $\mathbf{C}_0$ made of precisely two triangles in $\mathbf{T}_0(\tilde{z}; h)$ with a common hypotenuse, and the two hypotenuses are orthogonal each other. The cube between $\mathbf{C}_{-1}$ and $\mathbf{C}_0$ determined by two squares is subdivided into five tetrahedra with the manner shown in Fig. 1.3. All such cubes are treated in the same way.

Similarly, for $d \geq 0$, every square in $\mathbf{C}_d$ made of exactly two triangles in $\mathbf{T}_d$ and its four above-opposite squares in $\mathbf{C}_{d+1}$ determine an elementary guadrangular prism between $\mathbf{C}_d$ and $\mathbf{C}_{d+1}$, the prism is subdivided into fourteen tetrahedra as in Fig. 1.4

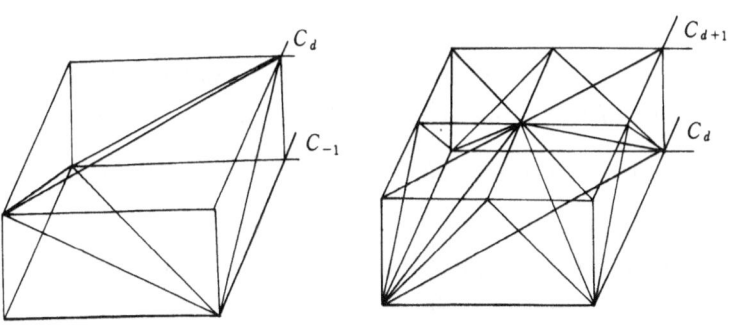

Figure 1.3 Figure 1.4

In this way we obtain the simplicial triangulation $\mathbf{T}(\tilde{z}; h)$ of the half-space $\mathbf{C} \times [-1, \infty)$, shortly denoted by $\mathbf{T}$. Notice that no new vertices have been added to define the triangulation, that is, all vertices of the triangulation $\mathbf{T}(\tilde{z}; h)$ are the vertices of the triangulation $\mathbf{T}_d(\tilde{z}; h)$ for $d = -1, 0, 1, 2, \cdots$. We denote the vertex set of $\mathbf{T}(\tilde{z}; h)$ by $V(\mathbf{T}(\tilde{z}; h))$ or $V(\mathbf{T})$.

In the algorithm to be presented in the next section, we shall be concerned very often with triples of points $\{(z_1, d_1), (z_2, d_2), (z_3, d_3)\}$ (or shortly $\{z_1, z_2, z_3\}$) which are the vertices of a triangular face of some tetrahedron in the triangulation. The name "triple" used in next several sections has always this meaning.

For the triangulation, the following lemma is obvious.

Lemma 1.2 *Suppose that* $\{(z_1, d_1), (z_2, d_2), (z_3, d_3)\}$ *is a triple in* $\mathbf{T}$. *Let* $d = \min\{d_1, d_2, d_3\}$. *Then* $d \leq d_k \leq d + 1$ *for* $k = 1, 2, 3$. ¶

In the case of Lemma 1.2, we say that the triple $\{z_1, z_2, z_3\}$ lies between levels $\mathbf{C}_d$ and $\mathbf{C}_{d+1}$. In particular, when $d_1 = d_2 = d_3$, we say that the triple $\{z_1, z_2, z_3\}$ lies in $\mathbf{C}_d$.

Let $\{(z_1, d_1), (z_2, d_2), (z_3, d_3)\}$ be a triple in $\mathbf{T}$. Define the diameter of the triple by

$$\mathrm{diam}\{(z_1, d_1), (z_2, d_2), (z_3, d_3)\} = \max\{|z_1 - z_2|, |z_2 - z_3|, |z_3 - z_1|\}.$$

One may instead denote it shortly by $\mathrm{diam}\{z_1, z_2, z_3\}$. Notice that it is, in fact, a projective diameter.

Lemma 1.3 *Suppose that the triple* $\{z_1, z_2, z_3\}$ *lies between levels* $\mathbf{C}_d$ *and* $\mathbf{C}_{d+1}$. *Then*
$$\mathrm{diam}\{z_1, z_2, z_3\} \leq \sqrt{2} h 2^{-d}.$$

Proof. This inequality is obvious since the diameter of all possible triples lying between levels $\mathbf{C}_d$ and $\mathbf{C}_{d+1}$ is easily found from Figs. 1.3 and 1.4. ¶

This lemma shows the fact that the higher the level $\mathbf{C}_d$ is, the smaller is the diameter of the triple in $\mathbf{C}_d$ or above $\mathbf{C}_d$.

Now, we turn to present the labelling rule for the vertices of the triangulation $\mathbf{T}(\tilde{z}; h)$.

Given any nonzero complex number $w = u + i v$, define $\arg w$ to be the unique real number α satisfying

$$-\pi < \alpha \leq \pi,$$

$$\cos \alpha = \frac{u}{\sqrt{u^2 + v^2}},$$

and

$$\sin \alpha = \frac{v}{\sqrt{u^2 + v^2}}.$$

Definition 1.4 Define an assignment $l : \mathbf{C} \to \{1, 2, 3\}$ by

$$l(z) = \begin{cases} 1, & \text{if} \quad -\pi/3 \ \le \arg f(z) \le \pi/3 \text{ or } f(z) = 0; \\ 2, & \text{if} \quad \pi/3 \ < \arg f(z) \le \pi; \\ 3, & \text{if} \quad -\pi \ < \arg f(z) < -\pi/3. \end{cases}$$

l is called the labelling of the z-plane $\mathbf{C}$ induced by the polynomial f, and $l(z)$ is called the label of z. See Fig. 1.5 for an illustration.

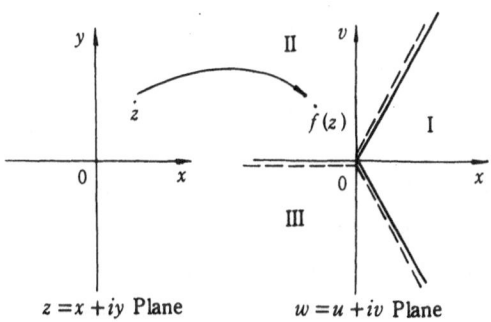

$z = x + iy$ Plane $w = u + iv$ Plane

Figure 1.5

Definition 1.5 Let $f_{-1}(z) = (z - \tilde{z})^n$, and $f_d(z) = f(z)$ for $d = 0, 1, 2, \cdots$. Define $l : V(\mathbf{T}(\tilde{z}; h)) \to \{1, 2, 3\}$ by

$$l(z, d) = \begin{cases} 1, & \text{if} \quad -\pi/3 \ \le \arg f_d(z) \le \pi/3 \text{ or } f_d(z) = 0; \\ 2, & \text{if} \quad \pi/3 \ < \arg f_d(z) \le \pi; \\ 3, & \text{if} \quad -\pi \ < \arg f_d(z) < -\pi/3. \end{cases}$$

l is called the labelling of $V(\mathbf{T}(\tilde{z}; h))$ induced by the polynomial f, and $l(z, d)$ is called the label of the vertex (z, d).

Notice that we use the same notation l in Definition 1.4 and in Definition 1.5. No confusions could arise since, for every vertex $(z, d) \in V(\mathbf{T}(\tilde{z}; h))$, it is clear that its label is either determined by $f(z)$ when $d \ge 0$ or determined by $(z - \tilde{z})^n$ when $d = -1$.

Based on the triangulation and the labelling, we can now introduce the concept of complete triples.

Definition 1.6   The triple $\{z_1, z_2, z_3\}$ is completely labelled by f if $\{l(z_1), l(z_2), l(z_3)\} = \{1, 2, 3\}$. In this case, we may also simply say that $\{z_1, z_2, z_3\}$ is a complete triple.

For simplicity of expressions and of specification, from now on we assume that $l(z_k) = k, k = 1, 2, 3$ for any given complete triple $\{z_1, z_2, z_3\}$.

The definition of complete triples does not point out whether the labels of its vertices are determined by $f(z)$ or by $(z - \tilde{z})^n$. In fact, for some triple $\{z_1, z_2, z_3\}$, it is possible that all of its vertices are labelled by $f(z)$, or all are labelled by $(z - \tilde{z})^n$, or partially labelled by $f(z)$ and partially labelled by $(z - \tilde{z})^n$.

The following result establishes certain connections between triples completely labelled by $f(z)$ and the zeros of $f(z)$.

Lemma 1.7   *Let $\{z_1, z_2, z_3\}$ be a complete triple whose labels are determined by $f(z)$, and $|f(z_k) - f(z_l)| \leq \eta$ for $k, l = 1, 2, 3$. Then for $k = 1, 2, 3$,*

$$|f(z_k)| \leq \frac{2\eta}{\sqrt{3}}.$$

Proof.   The sectors 1, 2 and 3 of the w-plane illustrated in Fig. 1.6 are the three ranges whose preimages are respectively labelled 1, 2 and 3. The label of a point z is determined by the sector into which $f(z)$ falls. According to the convention $l(z_k) = k$ for $k = 1, 2, 3$, if $|f(z_k) - f(z_l)| \leq \eta$ for $k, l = 1, 2, 3$, then $f(z_1)$ must lie in Sector 1 and both the distances between $f(z_1)$ and Sectors 2 and 3 are smaller than η. Thus $f(z_1)$ must lie in the shadowed area as shown in Fig. 1.6. Hence, $|f(z_1)| \leq 2\eta/\sqrt{3}$.

Similarly, $|f(z_2)| \leq 2\eta/\sqrt{3}$ and $|f(z_3)| \leq 2\eta/\sqrt{3}$. ¶

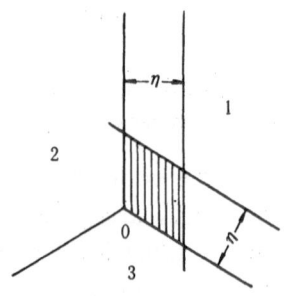

Figure 1.6

It is well-known that every polynomial function is uniformly continuous in any bounded domain. If we can find a triple completely labelled by f with small diameter, then the distances between the images of its vertices are also small and so are the distances between the images and the origin of the w-plane (by Lemma 1.7). If the distances are small enough, then every point of the triple can serve as a numerical zero of f with certain accuracy. Furthermore, we have learned that the higher the level $\mathbf{C}_d$ is, the smaller is the diameter of the triple in $\mathbf{C}_d$. These motivate to design an algorithm to locate complete triples whose projections always lie in a given bounded domain and the complete triples run through higher and higher levels. This will be done in the next section.

§2.   Complementary Pivoting Algorithm

We first present several useful lemmas.

Let $Q_m(\tilde{z}; h)$ denote the square in $\mathbf{C}$ with corners at $\tilde{z} + mh(\pm 1 \pm i)$, where m is a positive integer, that is, $Q_m(\tilde{z}; h)$ is a square with center at $\tilde{z}$, the lengths of its four sides are all $2mh$ and each side parallels the x-axis or the y-axis (Fig. 1.7).

A side of some triangle in the triangulation $\mathbf{T}$ is called an edge of the triangulation. The boundary $\partial Q_m(\tilde{z}; h)$ of $Q_m(\tilde{z}; h)$ is oriented counterclockwise. With the notation $\{z', z''\}$ for an edge on $\partial Q_m(\tilde{z}; h)$, we call $\{z', z''\}$ a positive edge if the direction of the edge from z' to z'' coincides with the direction of $\partial Q_m(\tilde{z}; h)$, otherwise $\{z', z''\}$ is called a negative edge. The triangles of $\mathbf{T}_{-1}(\tilde{z}; h)$ inside $Q_m(\tilde{z}; h)$ are oriented in the customary counterclockwise cyclic order of their vertices. With the notation $\{z', z'', z'''\}$ for a triangle in $\mathbf{T}_{-1}(\tilde{z}; h)$, we call $\{z', z'', z'''\}$ a positive triangle (triple) if the ordering of z', z'', z''' gives the positive direction of the triangle, otherwise it is a negative triangle. We may simply denote $Q_m(\tilde{z}; h)$ and $\partial Q_m(\tilde{z}; h)$ by Q_m and ∂Q_m respectively.

The angle spanned by z', z'' with respect to z^* is defined to be the unique angle no larger than π and spanned by two rays starting from z^* and passing through z' and z'' respectively. It is also called the angle spanned by the line segment $z'z''$ with respect to z^*.

By saying that an edge is labelled (k, l), we mean that its starting vertex is labelled k while the ending one is labelled l.

Lemma 2.1   *If* $m \geq \dfrac{3n}{2\pi}$, *then in the counterclockwise direction of*

∂Q_m, there are exactly n $(1,2)$-labelled edges on ∂Q_m and no $(2,1)$-labelled edges.

Proof. Let $\{z', z''\}$ be an edge on ∂Q_m. Denote by α the angle spanned by z' and z'' with respect to $\tilde{z}$. Refer to Fig. 1.7, it is evident that

$$0 < \alpha \leq \arctan\left(\frac{h}{mh}\right) < \frac{1}{m} \leq \frac{2\pi}{3n}.$$

Let β be the angle spanned by $w' = (z' - \tilde{z})^n$ and $w'' = (z'' - \tilde{z})^n$ with respect to the origin of the w-plane. Then

$$0 < \beta = n\alpha < \frac{n}{m} \leq \frac{2\pi}{3}.$$

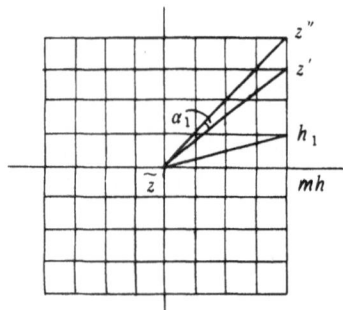

Figure 1.7

From the structure of Q_m and the properties of the function $w = (z - \tilde{z})^n$, the image of ∂Q_m runs around the origin of the w-plane n times. Since $0 < \beta < 2\pi/3$, the angle spanned by the images of some edge on ∂Q_m with respect to the origin is smaller than $2\pi/3$. Hence, starting from $w = (mh)^n$ in the w-plane, the image of ∂Q_m runs exactly n times from the sector 1 into the sector 2. Thus, in the z-plane, there are exactly n $(1,2)$-labelled edges on ∂Q_m.

Similarly, if $l(z') = 2$ then we have $l(z'') = 2$ or 3 and never $l(z'') = 1$ due to $0 < \beta < 2\pi/3$. Hence there is no $(2,1)$-labelled edges on ∂Q_m. ¶

Now, starting from $z = \tilde{z} + mh \in \partial Q_m$ and following ∂Q_m in its positive direction, we number the n $(1,2)$-labelled edges from 1 to n. This order of the $(1,2)$-labelled edges will be frequently used in the next sections.

Lemma 2.2 *If* $m \geq \dfrac{3(1+\sqrt{2})n}{4\pi}$, *then outside* Q_m *there is no complete triples labelled by* $(z - \tilde{z})^n$.

Proof. We first prove that if $z'z''$ is an edge on ∂Q_m or outside Q_m then the angle spanned by $z'z''$ with respect to $\tilde{z}$ is smaller than $\dfrac{2\pi}{3n}$. In fact, if $z'z''$ parallels the x-axis or the y-axis then the result follows directly from Lemma 2.1. Now, let $z'z''$ be the hypotenuse of a triangle in $\mathbf{T}_{-1}$ outside Q_m. Due to the structure of Q_m, the angle reaches its maximum only when the edge $z'z''$ intersecting with ∂Q_m. Without loss of generality, let k be a positive integer such that $z' = \tilde{z} + h(m+1+ik)$ and $z'' = \tilde{z} + h(m+i(k+1))$. Referring to Fig. 1.8, we have

$$\alpha = \arctan \frac{k+1}{m} - \arctan \frac{k}{m+1},$$

$$\tan \alpha = \frac{m+k+1}{m^2+m+k^2+k}.$$

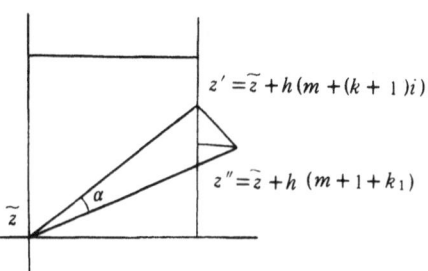

$z' = \tilde{z} + h(m+(k+1)i)$

$z'' = \tilde{z} + h(m+1+k_1)$

$\tilde{z}$

Figure 1.8

Hence, as a function of k, at $k = \sqrt{2m(m+1)} - m - 1$, $\tan \alpha$ takes its maximum

$$\frac{m+k+1}{m^2+m+k^2+k} = \frac{1}{2\sqrt{2m(m+1)} - 2m - 1}$$

$$< \frac{1}{2m(\sqrt{2}-1)}$$

$$= \frac{1+\sqrt{2}}{2m}.$$

Of course, the above inequality is true for any integer k. Notice also that $\alpha < \pi/2$, we obtain

$$\alpha < \tan \alpha < \frac{1+\sqrt{2}}{2m} \leq \frac{2\pi}{3n}.$$

Now let $\{z', z'', z'''\}$ be a triple in $\mathbf{T}_{-1}(\tilde{z}; h)$ outside Q_m. Then the angle spanned by each edge of the triple with respect to $\tilde{z}$ is less than $\dfrac{2\pi}{3n}$. Let $w' = (z' - \tilde{z})^n, w' = (z'' - \tilde{z})^n$ and $w''' = (z''' - \tilde{z})^n$. Then the angle spanned by a pair of w', w'', w''' with respect to the origin of the w-plane is less than $2\pi/3$. Combining with the following Lemma (2.3), $\{z', z'', z'''\}$ is not a triple completely labelled by $(z - \tilde{z})^n$. ¶

Lemma 2.3 *Let $\{z', z'', z'''\}$ be a triple in the z-plane, and let w', w'', w''' denote respectively the images of z', z'', z''' under mapping $w = (z-\tilde{z})^n$ or $w = f(z)$. If none of w', w'', w''' is zero and the angle spanned by any pair of w', w'', w''' with respect to the origin of the w-plane is less than $2\pi/3$, then the triple $\{z', z'', z'''\}$ is not complete.*

Proof.   Suppose otherwise that $\{z', z'', z'''\}$ is complete and

$$l(z') = 1, \quad l(z'') = 2, \text{ and } l(z''') = 3.$$

In the w-plane, let α, β, γ denote respectively the less-than-2π angle from $0w'''$ to $0w'$, from $0w'$ to $0w''$, and from $0w''$ to $0w'''$.(cf. Fig. 1.9) Then $\alpha > 0, \beta > 0, \gamma > 0$ and $\alpha + \beta + \gamma = 2\pi$.

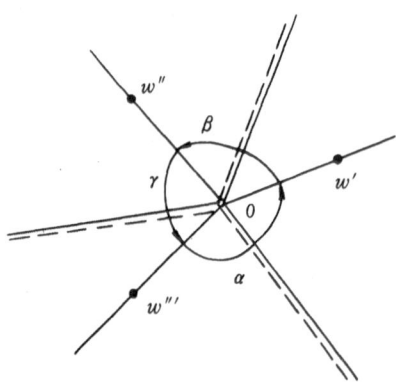

Figure 1.9

Now , if $\alpha > \pi$ then the angle spanned by $w'''w'$ with respect to the origin is $2\pi - \alpha$. By the assumption of this lemma, $2\pi - \alpha < 2\pi/3$, thus $\alpha > 4\pi/3$. But according to the labelling, $\alpha \leq 4\pi/3$. This is a contradiction. Similarly, $\beta > \pi$ or $\gamma > \pi$ will also lead to contradictions. Finally, if none of α, β, γ is larger than π, then α, β, γ are exactly the angles spanned respectively by the three pairs of w', w'' and w''' with respect to the origin, and thus are all smaller than $2\pi/3$. This contradicts

$\alpha + \beta + \gamma = 2\pi$ (cf. Fig. 1.9). This completes the proof. ¶

Next, we give the outline of the algorithm. Let

$$m = \left\lceil \frac{3(1+\sqrt{2})n}{4\pi} \right\rceil,$$

where $\lceil r \rceil$ stands for the least integer no less than r for any real number r. Set the square $Q_m(\tilde{z}; h)$ as above. Starting from each (1,2)-labelled edge on ∂Q_m, the algorithm is supposed to find complete triples. If the labels of a complete triple are not entirely determined by f, then the triple fail to offer enough information about a zero of f. But we will show that according to the algorithm to be presented next, the label of any vertex encountered must be determined by f whenever the computation left the level of $\mathbf{C}_{-1}$. In fact, this is true except for the vertices on level $\mathbf{C}_{-1}$. Thus, by Lemma 1.7, we can locate the zeros of f within a given accuracy.

Algorithm 2.4 The algorithm starts from the j-th (1,2)-labelled edge $\{z_1, z_2\}$ on $\partial Q_m, j = 1, \cdots, n$. Notice that the n edges are easy to be located.

Step 1 (Two-dimensional searching) If the storage of z_3 is not empty, go to Step 2. Otherwise, for the positive (1,2)-labelled edge $\{z_1, z_2\}$, there exists a unique vertex z' such that $\{z_1, z_2, z'\}$ is a positive triple in $\mathbf{T}_{-1}(\tilde{z}; h)$. Calculate the label $l = l(z')$ of z'. Set $z_l = z'$ and go to Step 1 (Hence, if $l = 3$ then the computation will enter three-dimensional searching from two-dimensional searching).

Step 2 (Dimension decreasing: from three-dimensional searching to two-dimensional searching) If $\{z_1, z_2, z_3\}$ is a negative triple in $\mathbf{T}_{-1}(\tilde{z}; h)$, delete z_3 from the storage, then we get a (1,2)-labelled edge $\{z_1, z_2\}$. Go to Step 1.

Step 3 (Three-dimensional searching) According to the triangulation, there is a unique vertex z' such that $\{z_1, z_2, z_3, z'\}$ determines a tetrahedron of $\mathbf{T}\{\tilde{z}; h\}$ and the order z_1, z_2, z_3, z' coincides the direction of a right-hand screw rotated in the cyclic order of z_1, z_2, z_3. Calculate $l = l(z')$ and set $z_l = z'$. Go to Step 2.

In the description of Algorithm 2.4, only Triangulations $\mathbf{T}_d$ and $\mathbf{T}$, and Definition 1.5 of the labelling are used. In Step 1, z' is determined by Triangulation $\mathbf{T}_d$, and $l(z')$ is determined by $(z' - \tilde{z})^n$ according to Definition 1.5. In Step 3, z' is found by Triangulation $\mathbf{T}$ and its label $l(z')$ is determined by $(z' - \tilde{z})^n$ when $d' = -1$ or by $f(z')$ when $d' \geq 0$.

After obtaining $l = l(z')$, set $z_l = z'$, that is, the vertex z_l with label l is replaced by a new vertex z' with the same label. We call this replacement the complementary pivoting procedure.

Analysis of entrances and exits

Now we present in details the Algorithm 2.4.

Starting from a (1,2)-labelled edge in Step 1, the algorithm generates first vertex z'; we say that the computation enters the triangle or the triple $\{z_1, z_2, z'\}$. Similarly, starting from a triple $\{z_1, z_2, z_3\}$ in Step 3, the algorithm finds z'; we say that the computation enters the tetrahedron $\{z_1, z_2, z_3, z'\}$.

In Step 1, if $l(z') = 1$ or 2, a new edge still labelled 1 and 2 is obtained and the computation of the algorithm enters a new triangle with two edges labelled 1 and 2. Imaging standing at the center of the triangle and reading the labels of the vertices counterclockwise, it is natural to call the (1,2)-labelled edge the entrance of the triangle and the (2,1)-labelled edge the exit. If $l(z') = 3$, a positive complete triple $\{z_1, z_2, z_3\}$ has been reached, and the computation enters a tetrahedron from the triple. In this case, we call the positive triangle or triple the exit of itself, besides the (1,2)-labelled edge its entrance.

In Step 2, in the case of dimension decreasing, a negative complete triple $\{z_1, z_2, z_3\}$ is the result of the previous computation. Now drop z_3, the computation leaves the negative triangle $\{z_1, z_2, z_3\}$ from its unique (2,1)-labelled edge. Thus we call the negative complete triple the entrance of itself while the (2,1)-labelled edge is its exit.

In the above cases, a (1,2)-labelled edge or a negative triangle $\{z_1, z_2, z_3\}$ is the entrance of the triple and a (2,1)-labelled edge or a positive triangle $\{z_1, z_2, z_3\}$ is the exit of the triple.(cf. Fig. 1.10)

In Step 3, the situation is simple since there is no dimension variations. Since a tetrahedron encountered has a known complete triple and two and only two of its four vertices have the same label, the tetrahedron has two complete triangular faces and the other two faces are not complete. After the same-label-substitution of the old one by the new one, a new complete triple is obtained. Imaging reading the labels of the vertices from the interior of the tetrahedron, a complete triangular face $\{z_1, z_2, z_3\}$ is positive if z_1, z_2, z_3 gives the counterclockwise ordering. Otherwise, $\{z_1, z_2, z_3\}$ is negative. Thus we call the positive complete triangular face or the triple the entrance of the tetrahedron and the negative complete triple its exit. (cf. Fig. 1.11)

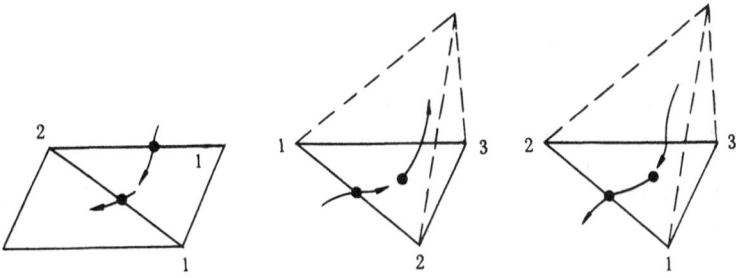

Figure 1.10

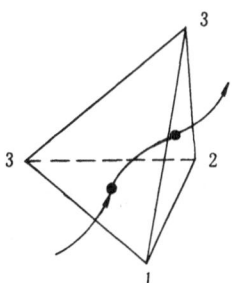

Figure 1.11

The entrances and the exits are collectively called the doors. Then we have the next lemma.

Lemma 2.5 *For any triple or tetrahedron whose vertices take labels from $\{1, 2, 3\}$, either it has no doors at all, or it has exactly a pair of doors: one is its entrance and the other is its exit.*

Proof. The concept of doors implies that any triple without label 1 or 2 doesn't have any doors. For a triple with label 1 and 2, if it has no label 3 then the (1,2)-labelled edge is its entrance while the (2,1)-labelled edge is its exit, and the third edge is certainly not a door. Therefore it has exactly a pair of doors. When the triple is complete, if it is negative then it is the entrance of itself and the (2,1)-labelled edge is its exit; If it is positive, then the (1,2)-labelled edge is its entrance and $\{z_1, z_2, z_3\}$ is the exit of itself. Whether it is negative or positive, the other two edges are with label 3, and so they are not doors. Hence it has just two doors.

Any tetrahedron without label 1, 2 or 3 has no doors. If a tetrahedron

contains all of the three labels 1, 2 and 3 then there are exactly a pair of
its vertices with a same label, thus a pair of its four faces are not complete
and the other two faces complete. In this case, one complete triple is its
entrance and the other its exit. This completes the proof. ¶

Based on the above discussions, we have the following proposition on
feasibility of the algorithm.

Lemma 2.6 *Starting from a $(1,2)$-labelled edge on ∂Q_m, the computation of Algorithm 2.4 will continue infinitely.*

Proof.   Since every triple or tetrahedron encountered has exactly one
entrance and one exit, the result is obvious. ¶

For a triple or tetrahedron, if the computation does somehow enter it
then the computation must leave it in the next step. In this case we may
say that the computation passes a triple or a tetrahedron.

§3.   Convergence, I

In this section we will prove that the computation of the algorithm starting from every (1,2)-labelled edge on ∂Q_m will converge to zero of the
polynomial f.

Notation 3.1 (*The computational sequence of simplices*)   For $j \in \{1, 2, \cdots, n\}$, the computation of the algorithm starting from the j-th
(1,2)-labelled edge on ∂Q_m passes a sequence of simplices, which are triples
(when two-dimensional searching) or tetrahedra(when three-dimensional
searching).   Let $\sigma_{j1}, \sigma_{j2}, \cdots, \sigma_{jn}, \cdots$ denote the sequence.   We call this
sequence the j-th computational sequence of simplices.   In this notation
σ_{jk} may be a triple or a tetrahedron.   If necessary, we will use σ_{jk}^2 or
$\dim(\sigma_{jk}) = 2$ to indicate that σ_{jk} is a triple while σ_{jk}^3 or $\dim(\sigma_{jk}) = 3$ to
indicate that σ_{jk} is a tetrahedron.

Notation 3.2 (*The computational sequence of vertices*)   Let σ_{jk} be
as in Notation 3.1. Let (z_{j1}, d_{j1}) be the vertex of σ_{j1} not lying on ∂Q_m.
For $k \geq 2$, if $\dim(\sigma_{jk}) \geq \dim(\sigma_{j,k-1})$ then let (z_{jk}, d_{jk}) be the vertex
of σ_{jk} but not a vertex of $\sigma_{j,k-1}$; If $\dim(\sigma_{jk}) < \dim(\sigma_{j,k-1})$ then let
$(z_{jk}, d_{jk}) = (z_{j,k-1}, d_{j,k-1})$.   Thus we obtain a sequence $\{(z_{jk}, d_{jk})\}_{k=1}^{\infty}$.
Call this sequence the j-th computational sequence of vertices. When we
are only concerned with the·projection of the sequence on the z-plane,
$\{z_{jk}\}_{k=1}^{\infty}$ is still called the j-th computational sequence of vertices.

It is important to note that $\dim(\sigma_{jk}) < \dim(\sigma_{j,k-1})$ can not occur successively in one sequence since only triples and tetrahedra are encountered. We will see in Corollary 3.6 that the case of $\dim(\sigma_{jk}) < \dim(\sigma_{j,k-1})$ can only occur a finite number of times. Thus the computation will call Step 2 only finite number of times.

The next lemma asserts that two-dimensional searching of the algorithm will be progressing only within Q_m.

Lemma 3.3   *If* $\dim(\sigma_{jk}) = 2$ *then* $\sigma_{jk} \subset Q_m$.

Proof.   Suppose otherwise that there are j, k such that $\dim(\sigma_{jk}) = 2$ and $\sigma_{jk} \not\subset Q_m$. Since two-dimensional searching only occurs in $\mathbf{C}_{-1}$, $\sigma_{jk} \subset \mathbf{C}_{-1}$. Without loss of generality, let σ_{jk} be the first triple lying outside Q_m in the sequence $\sigma_{j1}, \sigma_{j2}, \cdots$.

When $\dim(\sigma_{j,k-1}) = 2$, then $\sigma_{j,k-1} \subset Q_m$. Thus the common edge of σ_{jk} and $\sigma_{j,k-1}$ is on ∂Q_m and is of labels 1 and 2. If its labels are $(1,2)$ according to the counterclockwise direction of ∂Q_m, then it is the entrance of $\sigma_{j,k-1}$. This contradicts to the ordering of $\sigma_{j,k-1}, \sigma_{jk}$ in the sequence. But if its labels are $(2,1)$, this will contradict to Lemma 2.1.

When $\dim(\sigma_{j,k-1}) = 3$, according to Step 3 of the algorithm, σ_{jk} is a complete triple lying on $\mathbf{C}_{-1}$ but outside Q_m. This contradicts to Lemma 2.2. ¶

In the previous section, we proved that every triple in $\mathbf{T}_{-1}$ or every tetrahedron in $\mathbf{T}$ has either no doors or exactly a pair of doors with one door as its entrance and the other its exit. But whether a door is an entrance or an exit is relative to the concerned simplex, a triple or a tetrahedron. For instance, as shown in Fig. 1.12, a letter indicates a vertex while its subscript indicates the label of the vertex. Let us look first at the door B_1C_2. It is an exit of the triangle BAC but an entrance of the triangle BCD. Similarly, the door $B_1C_2D_3$ is an exit of itself where the computation leaves from two-dimensional searching to three-dimensional searching, so it is also an entrance of the tetrahedron $BCDE$. Similarly, the door $B_1E_2D_3$ is the exit of $BCDE$ and the entrance of $BEDF$.

For the convenience of description, if a door is an exit of some triple or tetrahedron, we say the triple or the tetrahedron is the upstream simplex of the door. If the door is an entrance of some triple or of some tetrahedron, we call the triangle or the tetrahedron the downstream simplex of the door.

Lemma 3.4   *For every door not on* ∂Q_m, *it has a unique upstream* *simplex and a unique downstream simplex.* ¶

The proof of the lemma is obvious and thus is omitted. Notice that for a door on ∂Q_m, it has a unique downstream simplex and has no upstream simplex. In fact, in this case, it is a beginning of the computation.

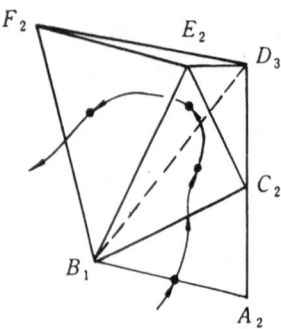

Figure 1.12

The next important lemma asserts that the algorithm passes any triple in $\mathbf{T}_{-1}$ or any tetrahedron in $\mathbf{T}$ at most one time.

Lemma 3.5   *If $(j, k) \neq (i, l)$ then $\sigma_{jk} \neq \sigma_{il}$.*

Proof.   Suppose otherwise that $\sigma_{jk} = \sigma_{il}$ and without loss of generality, suppose also $k \leq l$. According to Lemma 3.4, every door has a unique upstream simplex if it is not on ∂Q_m. Thus it is evident that $\sigma_{jk} = \sigma_{il}$ implies $\sigma_{j,k-1} = \sigma_{i,l-1}$ if $k > 1$. Finally we will have $\sigma_{j1} = \sigma_{i,l-k+1}$. Consider the door which is the entrance of σ_{j1}, it is the j-th (1,2)-labelled edge $\{z_1, z_2\}$ on ∂Q_m. If $l - k + 1 > 1$, i.e., $l > k$, then $\sigma_{i,l-k}$ is a triple with the edge as its exit and $\dim(\sigma_{i,l-k}) = 2$, but $\sigma_{i,l-k} \not\subset Q_m$. This leads to contradiction with Lemma 3.3. Hence $l = k$ and $\sigma_{j1} = \sigma_{i1}$. By Lemma 2.5, for the same simplex $\sigma_{j1} = \sigma_{i1}$, its entrance is unique. That is, the j-th (1,2)-labelled edge on ∂Q_m, the entrance of σ_{j1}, is just the i-th (1,2)-labelled edge on ∂Q_m, the entrance of σ_{i1}. Thus according to Notation 3.1, $i = j$. This contradicts $(j, k) \neq (i, l)$. ¶

Corollary 3.6   *For every $j \in \{1, 2, \cdots, n\}$, there is a positive integer K_j such that for any $k \geq K_j$, $\dim(\sigma_{jk}) = 3$.*

Proof.   Since two-dimensional searching occurs only within Q_m and the number of triangles in Q_m is finite, the result follows from Lemma 3.5.

¶

This corollary means that every computational sequence of simplices eventually consists of one tetrahedron except for the beginning part. So after a number of pivots the computation can't return to Q_m any more. Notice also that the inverse of Lemma 3.5 is not true although the triangles and the tetrahedra passed by the computation are all distinct. In fact, it is very possible that some vertices may appear several times in a computational sequence of vertices or even appear in different sequences. For example, as shown in Fig. 1.12, if instead of 2, the label of vertex E is 3, then the vertex A appears at least twice in the sequence. In fact, in this case, the corresponding piece of the computational sequence of vertices will be $\cdots, A_2, B_1, C_2, D_3, E_3, A_2, \cdots$.

Lemma 3.7 *There exists a constant R depending on $\tilde{z}, h$ and f such that no complete triples lies outside the cylinder $C(R) \times [-1, +\infty)$, where $C(R) = \{z : |z| \leq R\}$. In particular, the three-dimensional searching will not run out of $C(R) \times [-1, +\infty)$.*

Proof.   Set

$$R' = \max\left\{\frac{9\sqrt{2n}h}{2\pi} + |\tilde{z}|, \quad \left(\frac{9n}{2} + 1\right)|\tilde{z}|, \quad 9\max_{1\leq k\leq n}\frac{|a_k|}{2} + 1\right\},$$

and $R = R' + \sqrt{2}h$. Since $r = R' - |\tilde{z}| > 0$, for any

$$z \notin C(R') = \{z : |z| \leq R'\},$$

rewrite

$$
\begin{aligned}
f(z) &= z^n + a_1 z^{n-1} + \cdots + a_n \\
&= z^n\left(1 + \frac{a_1}{z} + \cdots + \frac{a_n}{z^n}\right) \\
&= (z - \tilde{z})^n\left(1 + \frac{\tilde{z}}{(z - \tilde{z})}\right)^n(1 + g(z)),
\end{aligned}
$$

where $g(z) = a_1/z + \cdots + a_n/z^n$. Suppose now that z', z'' are a pair of vertices of any triple lying outside $C(R) \times [-1, +\infty)$.(cf. Fig. 1.13) Let w', w'' be respectively the image of z' and of z'' under corresponding mappings $w = (z - \tilde{z})^n$ or $w = f(z)$. It is easy to verify that $w' \neq 0$ and $w'' \neq 0$. Furthermore, since the segment $z'z''$ lies outside of $C(R') \times [-1, +\infty)$, we have

$$
\begin{aligned}
\left|\arg\frac{w'}{w''}\right| \\
\leq\; &n\left|\arg\frac{z' - \tilde{z}}{z'' - \tilde{z}}\right| + n\left|\arg\frac{1 + \tilde{z}/(z' - \tilde{z})}{1 + \tilde{z}/(z'' - \tilde{z})}\right| + \left|\arg\frac{1 + g(z')}{1 + g(z'')}\right| \\
\leq\; &n\frac{\sqrt{2}h}{r} + n\frac{\pi}{2}\frac{|\tilde{z}|}{r} + 2\frac{\pi}{2}\left(\frac{|a_1|}{R'} + \cdots + \frac{|a_n|}{R'^n}\right).
\end{aligned}
$$

Notice that when both z' and z'' are labelled by $w = (z - \tilde{z})^n$, the second term and the third term can be deleted; When one of z' and z'' is labelled by $w = (z - \tilde{z})^n$ while the other is labelled by $w = f(z)$, the coefficient 2 in the second and the third terms of the right hand of the inequality can be omitted. But in any circumstances, the above inequality is true. Hence

$$\left|\arg\frac{w'}{w''}\right| < \frac{n\sqrt{2}h2\pi}{9\sqrt{2}nh} + \frac{n\pi|\tilde{z}|}{(9n/2)|\tilde{z}|} + \frac{2\max|a_k|}{9\max|a_k|/2} = \frac{2}{3}\pi.$$

Thus, Lemma 2.3 leads to obtain that any triple lying outside $C(R) \times [-1, +\infty)$ is not complete. ¶

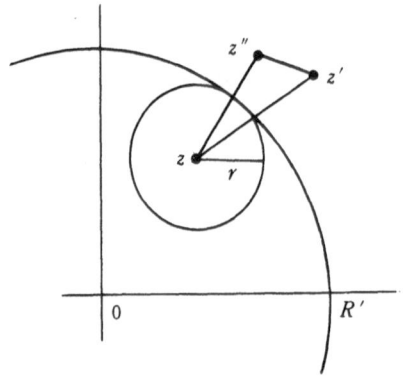

Figure 1.13

Now we are ready to prove the convergence of the algorithm.

Theorem 3.8 *The projection of every computational sequence of vertices on the z-plane has an accumulation point.*

Proof.   For any $j \in \{1, 2, \cdots, n\}, \{z_{jk}\}$ is an infinite sequence. Notice that $Q_m \subset C(R)$, by Lemmas 3.3 and 3.7, the two-dimensional searching and three-dimensional searching will never run out of $C(R) \times [-1, \infty)$. So $\{z_{jk}\} \subset C(R)$. But $C(R)$ is compact, hence the infinite sequence $\{z_{jk}\}$ must have an accumulation point in $C(R)$. ¶

Theorem 3.9 *If z^* is an accumulation point of the sequence $\{z_{jk}\}$, then $f(z^*) = 0$. In other words, every accumulation point of the sequence is a zero of the polynomial f.*

Proof.   Consider the corresponding computational sequence of simplices $\{\sigma_{jk}\}$ of $\{z_{jk}\}$. By Lemmas 3.5 and 2.6, the simplices in $\{\sigma_{jk}\}$ are all distinct. But according to the structure of $\mathbf{T}(\tilde{z}; h)$, the number of triangles lying in $Q_m \subset C_{-1}$ is finite, so is the number of the tetrahedra in $C(R) \times [-1, d]$ for any integer $d > 0$. Thus for any positive integer d there exists a positive integer $k(d)$ such that σ_{jk} must lie above the level $\mathbf{C}_d$ for every $k \geq k(d)$. By the manner of the determination of the sequence $\{(z_{jk}, d_{jk})\}$ in Notation 3.2, with no loss of generality we may assume that $d_{jk} \geq d$ for any $k \geq k(d)$.

Since z^* is an accumulation point of $\{z_{jk}\}$, there is a subsequence of $\{(z_{jk}, d_{jk})\}$ whose projective limit in the z-plane is z^*. Let $\{(z(k), d(k))\}$ denote the subsequence, then $\lim_{k \to \infty} z(k) = z^*$. No problem, we may assume also $d(k) \geq k + 1$.

Now to prove $f(z^*) = 0$, with the continuity of the polynomial, we need only to show that for any given number $\epsilon > 0$, there is a positive integer K such that $|f(z(k))| < \epsilon$ for all $k \geq K$. In fact, we may take

$$K = \left\lceil \log_2 \frac{2\sqrt{2}n^2 M R^{n-1} h}{\sqrt{3}\epsilon} \right\rceil,$$

where $M = \max\{1, |a_1|, \cdots, |a_{n-1}|\}$, R is as in Lemma 3.7. For $k \geq K$, let $\{z_1, z_2, z_3\}$ be a complete triple lying above the level $\mathbf{C}_K$ and containing $z(k)$ as a vertex, then

$$
\begin{aligned}
|f(z_1) - f(z_2)| \ &\leq \ |z_1^n - z_2^n| + |a_1||z_1^{n-1} - z_2^{n-1}| + \cdots + |a_{n-1}||z_1 - z_2| \\
&\leq \ \sqrt{2}h2^{-k}M(nR^{n-1} + (n-1)R^{n-2} + \cdots + 1) \\
&< \ \sqrt{2}h2^{-k}Mn^2 R^{n-1} \\
&\leq \ \frac{\sqrt{2}hMn^2 R^{n-1}\sqrt{3}\epsilon}{2\sqrt{2}n^2 M R^{n-1}h} \\
&= \ \frac{\sqrt{3}}{2}\epsilon.
\end{aligned}
$$

Similarly,

$$|f(z_1) - f(z_3)| < \frac{\sqrt{3}}{2}\epsilon,$$

$$|f(z_2) - f(z_3)| < \frac{\sqrt{3}}{2}\epsilon.$$

Hence, with Lemma 1.7 we obtain that

$$|f(z(k))| < \frac{2}{\sqrt{3}} \cdot \frac{\sqrt{3}\epsilon}{2} = \epsilon. \P$$

Notice that we did not assumed the existence of any zeros of the polynomial f. However, after presenting the algorithm we have now proved that projecting to the z-plane, each computational sequence of vertices produced by the algorithm has accumulation points and each of such accumulation points is a zero of the polynomial. Thereby, we do have constructively proved the next theorem.

Theorem 3.10 (Fundamental Theorem of Algebra) *Suppose that*

$$f(z) = z^n + a_1 z^{n-1} + \cdots + a_n,$$

where n is a positive integer, z the complex variable and $a_1, \cdots, a_n$ complex constants, then $f(z)$ does have zeros.

Based on the above discussions, we are now in position to express a polynomial in the form of factorization and to introduce the multiplicity of a zero of a polynomial.

Definition 3.11   Let $f(z) = z^n + a_1 z^{n-1} + \cdots + a_n = (z - \xi_1)^{k_1} \cdots (z - \xi_l)^{k_l}$, where l and $k_1, \cdots, k_l$ are all positive integers and $\xi_1, \cdots, \xi_l$ are distinct complex numbers. Call k_i the multiplicity of the zero ξ_i of the polynomial for $i = 1, \cdots, l$. In particular, if $k_i = 1$, ξ_i is called a simple zero of the polynomial; if $k_i > 1$, then ξ_i is called a multiple zero of the polynomial or a zero with multiplicity k_i.

Notice that a by-product of the factorization is $n = k_1 + \cdots + k_l$. Thus a polynomial of degree n has precisely n zeros, counting multiplicities.

§4.   Convergence, II

In this section we will further prove that every computational sequence of vertices produced by the algorithm will converge to one zero of the polynomial and the n computational sequences, starting from the n different $(1,2)$-labelled edges on ∂Q_m, will converge to the n zeros of the polynomial.

Theorem 4.1   *Every computational sequence of vertices converges exactly to one zero of the polynomial f.*

Proof.   In the previous section we have proved that every computational sequence $\{z_{jk}\}$ has at least one accumulation point z^* in the z-plane and the accumulation point is a zero of the polynomial. To prove $\lim_{k \to \infty} z_{jk} = z^*$, we need only to show that z^* is the unique accumulation point of the sequence in $C(R)$.

Suppose otherwise that z' is another accumulation point of the sequence in $C(R)$ and $z' \neq z^*$. Consider the domain

$$C' = \{z \in C(R) : |z - z'| \geq r', |z - z^*| \geq r'\},$$

where $r' = |z' - z^*|/4$, that is, C' is the domain $C(R)$ dug out two open disks with radius r' and centers at z' and z^* respectively. It is obvious that $C' \neq \emptyset$.

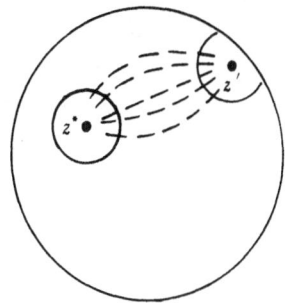

Figure 1.14

Since both z' and z^* are accumulation points of $\{z_{jk}\}$, for any integer K there are $k', k^* \geq K$ such that $|z_{jk'} - z'| < r'$ and $|z_{jk^*} - z^*| < r'$. Take K large enough such that $d_{jk} \geq \log_2(\sqrt{2}h/r') + 1$ and $|z_{jk} - z_{j,k-1}| \leq r'$ for all $k \geq K$. Without loss of generality, suppose $k' > k^*$. Since $|z' - z^*| > r'$, the part $\{z_{jk^*}, z_{j,k^*+1}, \cdots, z_{jk'}\}$ of the sequence $\{z_{jk}\}$ has a nonempty intersection with C'.

By the arbitrariness of K, $\{z_{jk}\} \cap C'$ is thus also an infinite sequence. Since C' is compact, so there is an accumulation point z'' of $\{z_{jk}\} \cap C'$ in C'. Of course, z'' is also an accumulation point of $\{z_{jk}\}$ and $z'' \neq z', z'' \neq z^*$.

Digging out again three disks with radius

$$r'' = \frac{1}{4} \min\{|z'' - z^*|, |z'' - z'|, |z' - z^*|\}$$

and centers at z', z'' and z^* from $C(R)$ and repeating the above discussion, we can get the fourth accumulation point of $\{z_{jk}\}$ different from z', z'' and z^*. Recursively, we will gain an infinite number of distinct accumulation points of $\{z_{jk}\}$ in $C(R)$. Then Theorem 3.9 would claim that the polynomial f has an infinite number of zeros in $C(R)$. This would leads to $f(z) \equiv 0$, a contradiction to the assumption that f is a polynomial of degree $n > 0$. The contradiction gives the theorem. ¶

In the rest of this section we will first prove that in the case of the polynomial with only simple zeros, the n computational sequences of vertices produced by the algorithm converge exactly to the n zeros of the polynomial in a one-to-one manner , then we will generalize the result to the case of arbitrary polynomials. For this purpose, now we introduce the following theorem.

Theorem 4.2 (Combinatorial Stokes' Theorem)   *Let Q be a bounded domain of the plane that can be triangulated into a finite number of triangles. Assign each triangle the positive direction of the plane, and assign each vertex of the triangulation a label from $\{1, 2, 3\}$ arbitrarily. Both the $(1, 2)$-labelled edges on ∂Q and the $(1, 3, 2)$-labelled triangles in Q are the entrances of the domain while the $(2, 1)$-labelled edges on ∂Q and the $(1, 2, 3)$-labelled triangles in Q are the exits of Q. Then the number of entrances of Q equals that of its exits.*

Proof.   By the assumptions, the number of triangles in Q and the number of edges on ∂Q are finite. Hence the number of exits and the number of entrances are also finite.

For an entrance of Q, if it is a (1,2)-labelled edge on ∂Q or a (1,3,2)-labelled triangle in Q, regard this edge or the (1,2)-labelled edge of this triangle as the start of Step 1 in Algorithm 2.4. Since the algorithm passes any triangle in Q at most one time, the algorithm will eventually reach either a (1,2,3)-labelled triangle in Q or a (2,1)-labelled edge on ∂Q. Hence starting from an entrance of Q, the algorithm ends at an exit of Q. Furthermore, by Lemmas 2.5 and 3.4, the complementary pivoting procedure is reversible. In other words, the pivoting is two-way feasible. Therefore the start and the end of the procedure uniquely determines each other. The proof is thus completed. ¶

Notice that Q_m in Algorithm 2.4 is a special case of Q in the previous theorem. We encourage the reader to give several such domains Q, triangulate them into triangles and assign arbitrarily a label 1,2 or 3 for each vertex, then count the numbers of their entrances and their exits. You will always obtain that the number of the entrances is equal to that of the exits. An example is shown in Fig. 1.15, where $\triangle$ denotes an entrance and $\rightarrow$ indicates an exit. Notice that arc triangles are also consistent with Combinatorial Stokes' Theorem.

From the discussion of Theorems 3.8, 3.9, and Theorem 3.10, we can rewrite

$$f(z) = \prod_{j=1}^{n}(z - \xi_j),$$

where $\xi_1, \cdots, \xi_n$ are the n zeros of the polynomial f.

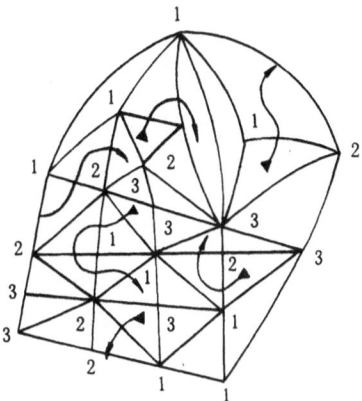

Figure 1.15

Lemma 4.3 *Suppose that ξ_j is a simple zero of the polynomial $f(z)$ for some $j \in \{1, 2, \cdots, n\}$. Then there exists at most one computational sequence of vertices produced by Algorithm 2.4 that converges to ξ_j.*

Proof.   In a small neighborhood of $z = \xi_j$,

$$f(z) = \prod_{j=1}^{n}(z - \xi_j) \approx (z - \xi_j) \prod_{i \neq j}(\xi_j - \xi_i),$$

that is, f is approximately a linear function in the neighborhood. Hence there is a disk with center at $z = \xi_j$ and a small radius (in the next chapter an estimate of the radius will be given in Lemma 2.3.5) such that for any z', z'' in the disk, the error caused by substituting $w = (z - \xi_j) \prod_{i \neq j}(\xi_j - \xi_i)$ for $w = f(z)$ to calculate $|\arg f(z')|$ is less than $\pi/24$; The error caused by substituting $w = (z - \xi_j) \prod_{i \neq j}(\xi_j - \xi_i)$ for $w = f(z)$ to calculate $|\arg f(z')/f(z'')|$ is less than $\pi/12$.

Given the plane $\mathbf{C}_d$ triangulated by $\mathbf{T}_d$, let τ be a triangle in which $z = \xi_j$ lies and $T(\tau)$ be the point-set convex hull of the triangles which have common vertices with τ (see Fig. 1.16). It is clear that we can always make $T(\tau)$ lie in the above-mentioned small disk for d large enough. Now, we will prove that there is uniquely a complete triple $\{z_1, z_2, z_3\}$ lying in the disk and the order of z_1, z_2, z_3 gives the positive direction of the triangle.

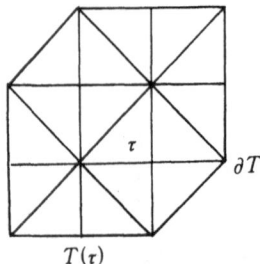

Figure 1.16

Let ∂T be the boundary of $T(\tau)$. It can be verified that the angle spanned by any edge on ∂T with respect to any point in τ is always in $(\pi/12, \pi/2)$. Therefore, when the image of ∂T under $w = (z - \xi_j) \prod_{i \neq j} (\xi_j - \xi_i)$ winds around the origin of the w-plane in its positive direction, for every edge $\{z', z''\}$ on ∂T, the angle spanned by the images of $\{z', z''\}$ under $w = (z - \xi_j) \prod_{i \neq j} (\xi_j - \xi_i)$ with respect to the origin lies in $(\pi/12, \pi/2)$. Thus the angle spanned by the images of $\{z', z''\}$ under $w = f(z)$ with respect to the origin lies between $\pi/12 - \pi/24 = \pi/24$ and $\pi/2 - \pi/24 = 13\pi/24$. Hence, according to the labelling, there exists exactly one (1,2)-labelled edge on ∂T and there is no (2,1)-labelled edges on ∂T. Applying the Combinatorial Stokes' Theorem to $T(\tau)$, we know that in $T(\tau)$ there is exactly one more positive complete triangle than the negative complete triangles .

Next, we prove that there is no negative complete triangles in the disk. Recall that, when defining the labelling in Definitions 1.4 and 1.5, we divide the w-plane into three fan-shaped sectors, each sector determining the label of the preimages of its points. Now consider the preimages of the three sectors in the above disk under $w = (z - \xi_j) \prod_{i \neq j} (\xi_j - \xi_i)$. Refer to Fig. 1.17, there $\Omega(1)$ indicates the fan-shaped sector whose points have label 1 while $\omega(1, 2)$ represents the fan-shaped sector whose points have either label 1 or label 2, and so on. Therefore, we can set the angle of the domain ω with respect to ξ_j to be $\pi/12$ and so the domain ω doesn't contain its boundary, and thus the angle of Ω is $7\pi/12$.

Let $\{z_1, z_2, z_3\}$ be a negative complete triangle in the disk. If at least two of its vertices lie in the Ωs(as the triangles A, B, and C in Fig. 1.17) then one of its inner angles is larger than $9\pi/12$. This contradicts the definition of the triangulation. In fact, the three inner angles of all mentioned triangles must be $\pi/2$, $\pi/4$ and $\pi/4$. When only one of its vertices lies in some Ω, if the other lies in some ω adjacent with the Ω and the third lies in the opposite Ω (as the triangle D in Fig. 1.17) then one of its angles is larger than $8\pi/12$; If the two other vertices lie respectively in the two ωs adjacent with the Ω (as the triangle E in Fig. 1.17) then one of its angle is larger than $7\pi/12$. These all contradict the definition of the triangulation. In the case of no vertices lying in the Ωs, if the three vertices lie in three different ωs then it is not a negative complete triple. Therefore, two of its vertices must lie in some ω.

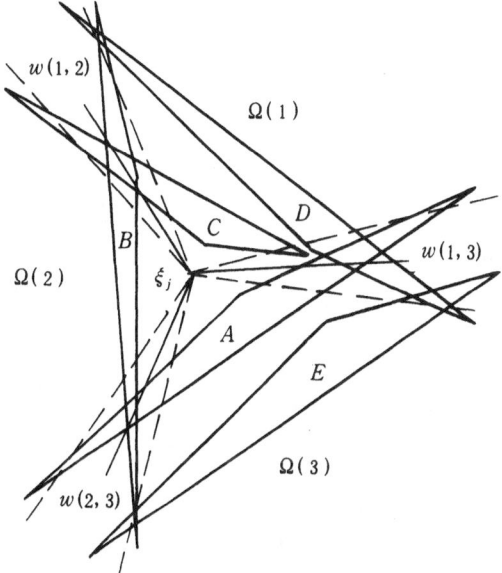

Figure 1.17

Thus it only remainds to discuss the case where two vertices lie in some ω and the third in the domain not adjacent to the ω. Without loss of generality, suppose $z_2, z_3 \in \omega(2,3), z_1 \in \omega(3,1) \cup \Omega(1) \cup \omega(1,2)$.

If $\xi_j \in \triangle z_1 z_2 z_3$ then $\angle z_1 < \pi/12$, this contradicts again the triangulation. Hence $\xi_j \notin \triangle z_1 z_2 z_3$. By the symmetry, we may suppose that z_2 and ξ_j are situated at different sides of $z_1 z_3$. Let z' denote the intersection of $z_1 z_3$ with the boundary of $\omega(3,1) \cup \Omega(1) \cup \omega(1,2)$. Consider the process of z_3 moving toward ξ_j along the ray through ξ_j and z_2 leaving ξ_j along the ray through ξ_j (see Fig. 1.18), this process enlarges the angle $\angle z_2 z' z_3$. It is easy to see that $\pi - 7\pi/12$ is an upper bound of the angle. Thus

$$\angle z_1 < \angle z_2 z' z_3 \le \pi - \frac{7\pi}{12} < \frac{\pi}{2}.$$

If $\angle z_3 = \pi/2$ then $z_2 z_3 = z_1 z_3 > \xi_j z_3$. Hence

$$\angle \xi_j z_2 z_3 < \alpha < \frac{\pi}{12}, \quad \angle z_1 z_3 \xi_j > \frac{\pi}{2} - \alpha - \frac{\pi}{12},$$

this leads to

$$\angle z_1 \xi_j z_3 + \angle \xi_j z_3 z_1 > \left(\frac{7\pi}{12} + \alpha\right) + \left(\frac{\pi}{2} - \alpha - \frac{\pi}{12}\right) = \pi,$$

a contradiction. If $\angle z_2 = \pi/2$ then

$$z_2 z_3 = z_1 z_2 > z_2 \xi_j$$

and thus

$$\angle z_2 z_3 \xi_j < \angle z_2 \xi_j z_3 < \frac{\pi}{12},$$

this leads to

$$\angle \xi_j z_2 z_3 > \pi - \frac{\pi}{12} - \frac{\pi}{12} > \frac{\pi}{2},$$

again a contradiction.

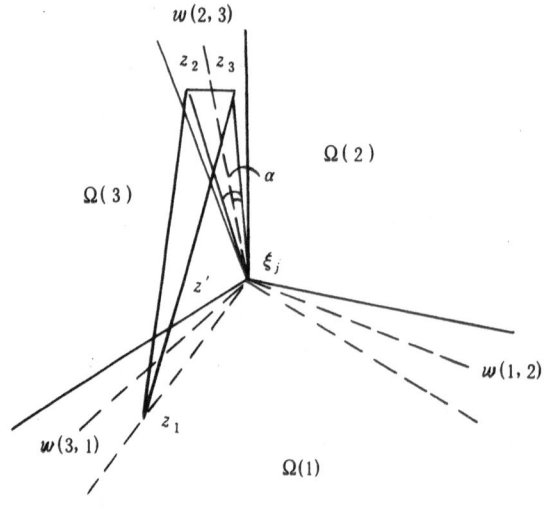

Figure 1.18

Therefore, there is no negative complete triangles in the disk. Thus the Combinatorial Stokes' Theorem guarantees that there is one and only one complete triangle lying in $T(\tau)$, and it is a positive one.

Finally suppose that $z'z''$ is an edge lying in the disk but outside $T(\tau)$. It is evident that the angle spanned by $z'z''$ with respect to $\xi_j \in \tau$ is less than or equal to $\pi/2$. Hence, the angle spanned by $f(z'), f(z'')$ with respect to the origin is less than or equal to $\pi/2 + \pi/12 < 2\pi/3$. Therefore, there is no complete triangles lying on the disk but outside $T(\tau)$

Now, we have proved that there is exactly one complete triangle lying in the small disk with center at ξ_j for d large enough and this complete triangle is positive. Thus at most one computational sequence of vertices can approach ξ_j infinitely, and so at most one computational sequence of vertices converges to ξ_j. ¶

Notice that the proof of the lemma eliminates the possibility of existence of any negative complete isosceles right triangle in the disk even when the triangle is not one of the triangulation.

Finally, we need a lemma concerning the dependence of the zeros on the coefficients

Lemma 4.4 *Every zero of a monic polynomial is continuous in the coefficients of the polynomial.*

Proof. Recall Definition 3.11, let $f(z) = z^n + a_1 z^{n-1} + \cdots + a_n = (z - \xi_1)^{k_1} \cdots (z - \xi_l)^{k_l}$, where l and $k_1, \cdots, k_l$ are positive integers with $k_1 + \cdots + k_l = n$, and $\xi_1, \cdots, \xi_l$ are all distinct.

For any $j \in \{1, 2, \cdots, k_l\}$, there exists an $\epsilon > 0$ such that ξ_j is the unique zero of the polynomial in $\{z : |z - \xi_j| \leq \epsilon\}$. With the argument principle of complex functions (for example, see [Rudin, 1974]), we have

$$k_j = \frac{1}{2\pi i} \int_{|z - \xi_j| = \epsilon} \frac{f'(z)}{f(z)} \, dz.$$

Notice that the left hand of the above equality is an integer while the right is a continuous function in the coefficients $a_1, \cdots, a_n$. Thus the integral keeps constant under small perturbations of $a_1, \cdots, a_n$, and so does the number of zeros of the polynomial in the small disk $\{z : |z - \xi_j| \leq \epsilon\}$. By arbitrariness of the small ϵ, the result follows. ¶

This lemma can be naturally extended to the case of all polynomials. Let $f(z) = a_0 z^n + a_1 z^{n-1} + \cdots + a_n = a_0 (z - \xi_1)^{k_1} \cdots (z - \xi_l)^{k_l}$, where l and $k_1, \cdots, k_l$ are positive integers with $k_1 + \cdots + k_l = n$, and $\xi_1, \cdots, \xi_l$ are all distinct. Then counting multiplicities, the polynomial $g(z) = a_0' z^n + a_1' z^{n-1} + \cdots + a_n'$ has exactly k_j zeros in a small disk with center at ξ_j as long as $(a_0', \cdots, a_n')$ approaches $(a_0, \cdots, a_n)$ enough, $j = 1, 2, \cdots, l$. In this sense we have

Lemma 4.5 *The zeros of a polynomial are continuous in the coefficients of the polynomial.* ¶

Now we are ready to present our main result.

Theorem 4.6   *The n computational sequences of vertices converge to n zeros of the polynomial. Furthermore, there are exactly μ computational sequences of vertices which will converge to zero if the multiplicity of the zero is μ.*

Proof.   When the polynomial has no multiple zeros, the conclusion is quite straightforward. In fact, in this case, Theorem 4.1 says that every computational sequence converges exactly to a zero of the polynomial, while Lemma 4.3 asserts that for every zero of the polynomial, there is at most one computational sequence which converges to zero. Thereby, every zero of the polynomial is a limit of some computational sequence. Remember that there are exactly n sequences for a polynomial of degree n. This is a one-to-one convergence.

Now we turn to considering the case when the polynomial has some multiple zeros. Notice that for any given $\epsilon > 0$, the derivative of $f(z) + \epsilon$ is still $f'(z)$, a polynomial of degree $n - 1$. Let $\xi'_1, \cdots, \xi'_{n-1}$ be the zeros of $f'(z)$. Then $f(z) + \epsilon$ has no multiple zeros for any $\epsilon \neq -f(\xi'_j), j = 1, 2, \cdots, n - 1$. Therefore, we can choose $\epsilon_0 > 0$ such that for any $0 < \epsilon \leq \epsilon_0$, $f(z) + \epsilon$ has no multiple zeros. Applying the algorithm to $f(z) + \epsilon$, we obtain n computational sequences $\{z_{jk}(\epsilon)\}, j = 1, \cdots, n, k = 1, 2, \cdots$. The n sequences converge to the all n simple zeros of $f(z) + \epsilon$ since it is a polynomial with only simple zeros. Without loss of generality, assume $\lim_{k \to \infty} z_{jk}(\epsilon) = \xi_j(\epsilon), j = 1, \cdots, n$, where $\xi_1(\epsilon), \cdots, \xi_n(\epsilon)$ are the n simple zeros of $f(z) + \epsilon$

For the relationship of $\{z_{jk}(\epsilon)\}$ and $\{z_{jk}\}$, we assert that for a given integer K there exists an $\epsilon_1 \in (0, \epsilon_0]$ such that $z_{jk}(\epsilon) = z_{jk}$ for any $1 \leq k \leq K$ and $0 < \epsilon < \epsilon_1$. In fact, for any point z, the label of z determined by $f(z)$ is equal to the one of z determined by $f(z) + \epsilon$ if ϵ is small enough. This is because the three labelling sectors of the w-plane are all open in their right boundaries (Fig. 1.5). Thus there exists an $\epsilon_1 \in (0, \epsilon_0]$ for the finite K such that for all $\epsilon \in (0, \epsilon_1]$, both $f(z)$ and $f(z) + \epsilon$ assign same labels on $z_{j1}, \cdots, z_{jK}$. Therefore, with the arbitrariness of K, we have $\lim_{k \to \infty} \lim_{\epsilon \to 0^+} z_{jk}(\epsilon) = \lim_{k \to \infty} z_{jk}$.

Based on the proof of Lemma 4.4, we have furthermore that $\xi_j = \lim_{\epsilon \to 0^+} \xi_j(\epsilon), j = 1, 2, \cdots, n$, where $\xi_1, \cdots, \xi_n$ are the zeros of $f(z)$. Next we will prove that $\xi_j = \lim_{k \to \infty} z_{jk}, j = 1, 2, \cdots, n$.

A natural way to do this is to verify

$$\xi_j = \lim_{\epsilon \to 0+} \xi_j(\epsilon) = \lim_{\epsilon \to 0+} \lim_{k \to \infty} z_{jk}(\epsilon) = \lim_{k \to \infty} \lim_{\epsilon \to 0+} z_{jk}(\epsilon) = \lim_{k \to \infty} z_{jk}.$$

But it is quite difficult to establish the interchangeability of the double limits. However, this idea motivates to prove indirectly the result by constructing a uniformly convergent subsequence of the sequence.

For any j and $\epsilon, 1 \le j \le n$ and $0 < \epsilon \le \epsilon_0$, let $\tilde{z}_{jd}$ and $\tilde{z}_{jd}(\epsilon)$ be respectively the first vertex in $\{z_{jd}\}$ and $\{z_{jd}(\epsilon)\}$ reaching the levels $\mathbf{C}_d, d = 0, 1, 2, \cdots$. For these subsequence we have

$$\lim_{d \to \infty} \tilde{z}_{jd} = \lim_{k \to \infty} z_{jk},$$

$$\lim_{d \to \infty} \tilde{z}_{jd}(\epsilon) = \lim_{k \to \infty} z_{jk}(\epsilon) = \xi_j(\epsilon)$$

and

$$\xi_j = \lim_{\epsilon \to 0+} \xi_j(\epsilon) = \lim_{\epsilon \to 0+} \lim_{d \to \infty} \tilde{z}_{jd}(\epsilon) = \lim_{d \to \infty} \lim_{\epsilon \to 0+} \tilde{z}_{jd}(\epsilon) = \lim_{d \to \infty} \tilde{z}_{jd}.$$

Here the interchange of the double limits is not problematic. In fact, no matter how small $\epsilon \in (0, \epsilon_0]$ is , the subsequence $\{\tilde{z}_{jd}(\epsilon)\}$ does always reach the level $\mathbf{C}_d$ in d steps, so $\{\tilde{z}_{jd}(\epsilon)\}$ is uniformly convergent for any $\epsilon \in (0, \epsilon_0]$, that is, given $\eta > 0$, we can find a $D > -\log_2(\eta/((1 + 3h/4)\sqrt{2}h))$ (Lemma 2.1.4* in the next chapter will provide a concrete estimate) such that for any $d \ge D$ and any $\epsilon \in (0, \epsilon_0]$,

$$|\tilde{z}_{jd}(\epsilon) - \xi_j(\epsilon)| < (1 + 3n/4)\sqrt{2}h2^{-d} \le (1 + 3n/4)\sqrt{2}h2^{-D} < \eta.$$

Thus we have proved that

$$\xi_j = \lim_{d \to \infty} \tilde{z}_{jd} = \lim_{k \to \infty} z_{jk}, \quad j = 1, \cdots, n.$$

Now suppose ξ is a zero of $f(z)$ with multiplicity μ. By virtue of the argument principle and on the the choice of ϵ_0, we have

$$\mu = \frac{1}{2\pi i} \int_{|z-\xi|=\epsilon_0} \frac{f'(z)}{f(z)} \, dz.$$

But $(f(z) + \epsilon)' = f'(z)$. The difference between $(f(z) + \epsilon)'/(f(z) + \epsilon)$ and $f'(z)/f(z)$ is uniformly very small on the circle $|z - \xi| = \epsilon_0$ for ϵ_0 small enough. Thus we have

---

*By Lemma 2.1.4 we mean Lemma 1.4 in Chapter 2.

$$\frac{1}{2\pi i} \int_{|z-\xi_j|=\epsilon} \frac{(f(z)+\epsilon)'}{(f(z)+\epsilon)} \, dz = \mu$$

for ϵ small enough since as an integer the integral keeps constants under small perturbations. The zeros of $f(z) + \epsilon$ are all simple, thus there are exactly μ computational sequences which converge to μ zeros of $f(z) + \epsilon$ in the disk $\{z : |z-\xi| < \epsilon_0\}$. Finally, with the arbitrariness of small $\epsilon_0 > 0$ and the existence of the corresponding $\epsilon_1 \in (0, \epsilon_0]$ for any integer K such that $z_{jk}(\epsilon) = z_{jk}$ for all $k = 1, \cdots, K$ and $0 < \epsilon \leq \epsilon_1$, there are exactly μ computational sequences which converge to ξ. ¶

Notice that in the above discussions, the ordering of the computational sequences is same as that of their starting (1,2)-labelled edges on ∂Q_m in its counterclockwise direction. To be more specific, we only declare that the counting will start at the center of the east boundary of Q_m. In the next chapter we will assign for the n zeros of the polynomial the same ordering as that of the corresponding computational sequences.

Chapter 2

Efficiency of Kuhn's Algorithm

In this chapter we will discuss the problems concerning the efficiency of
Kuhn's algorithm such as the computational complexity and the monotonicity of the computations.

In Section 1 we will establish an error estimate for Kuhn's algorithm
which is essential to the computational complexity. In Section 2 we will
prove that the cost of computing all zeros of a polynomial of degree n to
an accuracy of ϵ grows no faster than $O(n^3\log_2(n/\epsilon))$ provided $\epsilon > 0$ is
small enough. Thus in the view of the computational complexity theory,
Kuhn's algorithm is a polynomial-time algorithm, and a fairly efficient
algorithm.

On the other hand, the monotonicity problem of its computations
is also an important factor which deeply influences the efficiency of the
algorithm. So in Sections 3 and 4 we will discuss the monotonicity of
Kuhn's algorithm in two different ways, and the result obtained is that the
computations of Kuhn's algorithm are eventually monotonous except some
starting parts. These are also useful for us to understand the algorithm.

§1. Error Estimate

Roughly speaking, there are two kinds of accuracy criteria for computing
the solutions of an equation $f(z) = 0$. If the accuracy demand $\epsilon > 0$, there
exists either a point z such that $|z - \xi| < \epsilon$ for some point ξ with $f(\xi) = 0$,
or a point z with $|f(z)| < \epsilon$. In both cases, z is called a numerical solution
of $f(z) = 0$.

Recall that the proof of Theorem 1.3.9 has already provided Kuhn's
algorithm an error estimate of the second criterion $|f(z)| < \epsilon$. However,

the estimate is rather rough, and it depends not only on the degree but also on the coefficients of the polynomial.

Now we present an error estimate of the first criterion $|z - \xi| < \epsilon$. This estimate will be independent of the concrete coefficients of the polynomial, and so it will be a uniform error-estimate for the polynomials with same degree.

For 1, we always assume in this chapter that $f(z)$, the polynomial concerned, is a monic polynomial of degree n with complex coefficients. Besides, for any complex number z, its argument is always restricted to $\arg(z) \in (-\pi, \pi]$.

Lemma 1.1   *For any complex number w, if $|w| < 1$ then*

$$| \arg(1 + w)| \leq \frac{\pi}{2}|w|.$$

Proof.   It is well-known that $\lambda \leq \dfrac{\pi}{2} \sin \lambda$ for all $0 \leq \lambda \leq \pi/2$. Hence, if $|w| < 1$ then

$$|\arg(1 + w)| \leq \arcsin|w| \leq \frac{\pi}{2}\sin(\arcsin|w|) = \frac{\pi}{2}|w|.\P$$

Theorem 1.2   *If $\{z_1, z_2, z_3\}$ is a triangle completely labelled by f then the distance between the triangle and some zero of f is not greater than $3n\delta/4$, where $\delta = \operatorname{diam}\{z_1, z_2, z_3\}$ is the projective diameter of the triangle , and n is the degree of the polynomial f.*

Proof.   The case $n = 1$ is trivial. Now let $n > 1$. Rewrite $f(z) = \prod_{j=1}^{n}(z - \xi_j)$, where $\xi_1, \cdots, \xi_n$ are the n zeros of f. Suppose that the theorem is not true, that is, $|z_k - \xi_j| > 3n\delta/4$ for all $k = 1, 2, 3$ and $j = 1, \cdots, n$. Then for all $j = 1, \cdots, n$,

$$\left|\frac{z_2 - z_1}{z_1 - \xi_j}\right| < \frac{\delta}{3n\delta/4} = \frac{4}{3n} < 1.$$

By Lemma 1.1, we obtain that

$$\left|\arg\frac{f(z_2)}{f(z_1)}\right| = \left|\arg\frac{(z_2 - \xi_1) \cdots (z_2 - \xi_n)}{(z_1 - \xi_1) \cdots (z_1 - \xi_n)}\right|$$

$$\leq \sum_{j=1}^{n} \left|\arg\frac{(z_2 - \xi_j)}{(z_1 - \xi_j)}\right|$$

$$= \sum_{j=1}^{n} \left| \arg\left(1 + \frac{(z_2 - z_1)}{(z_1 - \xi_j)}\right) \right|$$

$$\leq \sum_{j=1}^{n} \frac{\pi}{2} \left| \frac{(z_2 - z_1)}{(z_1 - \xi_j)} \right|$$

$$< n \cdot \frac{\pi}{2} \cdot \frac{4}{3n} = \frac{2\pi}{3}. \tag{I}$$

Similar arguments lead to

$$\left| \arg\frac{f(z_1)}{f(z_3)} \right| \leq \frac{2\pi}{3} \tag{II}$$

and

$$\left| \arg\frac{f(z_2)}{f(z_3)} \right| \leq \frac{2\pi}{3}. \tag{III}$$

However, the labelling implies that the following three inequalities

$$\arg f(z_2) - \arg f(z_1) > \frac{4\pi}{3},$$

$$\arg f(z_1) - \arg f(z_3) > \frac{4\pi}{3}$$

and

$$\arg f(z_2) - \arg f(z_3) < \frac{2\pi}{3}$$

are all impossible. Therefore, (I), (II) and (III) are respectively equivalent to

$$0 < \arg f(z_2) - \arg f(z_1) < \frac{2\pi}{3}, \tag{IV}$$

$$0 < \arg f(z_1) - \arg f(z_3) < \frac{2\pi}{3} \tag{V}$$

and

$$\arg f(z_2) - \arg f(z_3) > \frac{4\pi}{3}. \tag{VI}$$

Then subtracting (IV) from (VI) gives

$$\arg f(z_1) - \arg f(z_3) > \frac{2\pi}{3},$$

which contradicts to the inequality (V). The proof is completed. ¶

The following useful lemma is directly from the proof of the previous theorem.

Lemma 1.3 *If $|\arg(f(z')/f(z''))| < 2\pi/3$ for every pair of vertices of some triangle then the triangle is not completely labelled by f.* ¶

In fact, Lemma 1.3 is a variant of Lemma 1.2.3. If the angle , spanned by the images of any pair of vertices of the triangle with respect to the origin of w-plane, is less than $2\pi/3$, then the triangle is not complete unless one of the vertices is already a zero of f.

Now return to the algorithm. We have

Lemma 1.4 *Give the accuracy demand $\epsilon > 0$. Let*

$$D = \lceil \log_2(\sqrt{2}h(1 + 0.75n)/\epsilon) \rceil.$$

To compute the zeros of the polynomial, no computations over $\mathbf{C}_D$ are necessary. In fact, if $\{z_1, z_2, z_3\}$ is a complete triangle lying in $\mathbf{C}_D$, the distance between any of z_1, z_2, z_3 and some zero of f is not greater than ϵ.

Proof. Let δ be the diameter of triangles in $\mathbf{C}_D$. By Theorem 1.2, the distance between $\{z_1, z_2, z_3\}$ and some zero of f is less than $3n\delta/4$. Therefore, the distance between any vertex of the triangle and the zero is less than $\delta + 3n\delta/4 = (1 + 0.75n)\delta$.

On the other hand, for a triangle in $\mathbf{C}_D$, its diameter

$$\delta = \sqrt{2}h2^{-D} \leq \frac{\sqrt{2}h}{\sqrt{2}h(1 + 0.75n)/\epsilon} = \frac{\epsilon}{1 + 0.75n}.$$

Hence, for any complete triangle in $\mathbf{C}_D$, the distances between the vertices of the triangle and some zeros of the polynomial are not greater than ϵ, that is, the computations have provided us some numerical zeros within the accuracy demand. ¶

Combinning the computational accuracy criterion with Algorithm 2.4 in Chapter 1, we have

Algorithm 1.5

Set $D = \lceil \log_2(\sqrt{2}h(1 + 0.75n)/\epsilon) \rceil$ and $j := 1$.

Step 0 If $j = n + 1$, stop. Otherwise, let $\{z_1, z_2\}$ be the j-th (1,2)-labelled edge on ∂Q_m.

Step 1 (two-dimensional searching)   If the storage of z_3 is not empty, go to Step 2. Otherwise, let z^i be the unique vertex such that $\{z_1, z_2, z'\}$ represents a positive triangle in $\mathbf{T}_{-1}(\tilde{z}; h)$. Calculate $l := l(z')$. If $l = 1$ or 2 then $z_l := z'$ and return Step 1. If $l(z') = 3$ then let $z_3 := z'$ and go to Step 2.

Step 2 (dimension decreasing)   If $\{z_1, z_2, z_3\}$ is a negative triangle in $\mathbf{T}_{-1}(\tilde{z}; h)$ then delete z_3 from the storage and go to Step 1. Otherwise, go to Step 3.

Step 3 (three-dimensional searching)   Let z' be the unique vertex of the triangulation $T(\tilde{z}; h)$ such that $\{z_1, z_2, z_3, z'\}$ is a tetrahedron of the triangulation and z' is indicated by the direction of a right-handed screw rotated in the cyclic order z_1, z_2, z_3. Calculate $l := l(z')$ and let $z_l := z'$. If $d_1 + d_2 + d_3 = 3D$, print $\xi_j := z_1$ as the j-th numerical zero of the polynomial, then let $j := j + 1$ and go to Step 1; if $d_1 + d_2 + d_3 = -3$, then go to Step 2; otherwise return Step 3.

Kuhn's algorithm may even be applied to compute solutions of transcendental equations (refer to [Wang, 1986] for some discussions). In this case, of course, on principle the convergence needs additional analysis.

§2.   Cost Estimate

Generally speaking, an algorithm is usually designed to deal with a special kind of problems, and the cost of its implementation and/or its computations grows as the "size" of the problems increases. If the cost grows exponentially with the "size", the algorithm is considered to be unacceptable. On the contrary, if the cost is bounded by a polynomial in the "size", the algorithm is regarded to be practically tractable.

In the case of finding zeros of polynomials, the degree of the polynomial is a natural size of the problem. In this section we will prove that the cost of Kuhn's algorithm for finding all zeros of a polynomial of degree n to an accuracy of ϵ grows no faster than $O(n^3 \log_2(n/\epsilon))$ as n increases if $\epsilon > 0$ is small enough.

The cost of computing solutions of some function equation is usually measured by the number of evaluations of the function. But for Kuhn's algorithm, it is known that the number of evaluations of the polynomial is not greater than the number of the tetrahedra encountered by the computation. Therefore, to obtain an upper bound of the cost, it suffices to give an upper bound of the number of the tetrahedra encountered.

For the simplicity of discussions, the triangulation in Kuhn's algorithm is specified to be $\mathbf{T}(0;1)$, that is, we always have $\tilde{z} = 0$ and $h = 1$.

Let

$$M = \sqrt{2} + \max\left\{\frac{3\sqrt{2}(2+\pi)n}{4\pi}, 1 + \frac{5}{4} \cdot \frac{n}{n-1} \max|a_j|\right\},$$

and

$$\Lambda = \{z : |z| < M\} \times [-1, +\infty).$$

Lemma 2.1 $\sigma_{jk} \subset \Lambda$ for $j = 1, \cdots, n$ and $k = 1, 2, \cdots$, i.e., the computations of Kuhn's algorithm are entirely proceeded within the big circular cylinder Λ.

Proof.   It is clear that $Q_m \subset \Lambda$. Hence, if $\dim(\sigma_{jk}) = 2$ then Lemma 1.3.3 implies $\sigma_{jk} \subset \Lambda$. Now we need only discuss the case of $\dim(\sigma_{jk}) = 3$.

Since the tetrahedron σ_{jk} has always a pair of complete triangular faces and it is exactly spanned by the pair of triangles, it suffices to show that every complete triangle must lie in Λ. Let

$$r = M - \sqrt{2} \geq \max\left\{\frac{3\sqrt{2}(2+\pi)n}{4\pi}, 1 + \frac{5}{4} \cdot \frac{n}{n-1} \max|a_j|\right\},$$

and

$$\Lambda' = \{z : |z| < r\} \times [-1, +\infty).$$

Since the diameter of any triangle in $\mathbf{T}(0;1)$ is not greater than $\sqrt{2}$, by Lemma 1.4, we need only to show that for every edge $\{z', z''\}$ outside the circular cylinder Λ',

$$\left|\arg \frac{f_d(z')}{f_d(z'')}\right| < \frac{2\pi}{3},$$

where $f_d(z) = z^n$ for $z \in \mathbf{C}_{-1}$, otherwise $f_d(z) = f(z)$.

Rewrite

$$f(z) = z^n\left(1 + \frac{a_1}{z} + \cdots + \frac{a_n}{z^n}\right) = z^n(1 + g(z)).$$

If $z' \notin \mathbf{C}_{-1}$ and $z'' \notin \mathbf{C}_{-1}$ then

$$|g(z'')| \leq \frac{|a_1|}{r} + \cdots + \frac{|a_n|}{r^n} \leq \frac{\max|a_j|}{r-1} \leq \frac{n-1}{n},$$

$$|g(z') - g(z'')| \leq |a_1| \left| \frac{1}{z'} - \frac{1}{z''} \right| + \cdots + |a_n| \left| \frac{1}{z'^n} - \frac{1}{z''^n} \right|$$

$$\leq \max |a_j| \cdot |z' - z''| \left(\frac{1}{r^2} + \frac{2}{r^3} + \cdots + \frac{n}{r^{n+1}} \right)$$

$$\leq \frac{\sqrt{2} \max |a_j|}{(r-1)^2}$$

$$\leq \frac{\sqrt{2}}{r-1} \cdot \frac{n-1}{n}$$

$$\leq \frac{n-1}{n} \sqrt{2} \bigg/ \left(\frac{3\sqrt{2}(2+\pi)n}{4\pi} - 1 \right)$$

$$< \frac{n-1}{n} \sqrt{2} \bigg/ \frac{3\sqrt{2}(2+\pi)(n-1)}{4\pi}$$

$$= \frac{4\pi}{3(2+\pi)n},$$

$$\left| \frac{g(z') - g(z'')}{1 + g(z'')} \right| < \frac{4\pi}{3(2+\pi)n} \bigg/ \left(1 - \frac{n-1}{n} \right) = \frac{4\pi}{3(2+\pi)} < 1.$$

With Lemma 1.1 we obtain that

$$\left| \arg \frac{f_d(z')}{f_d(z'')} \right| = \left| \arg \frac{f(z')}{f(z'')} \right|$$

$$\leq n \left| \arg \frac{z'}{z''} \right| + \left| \arg \left(1 + \frac{g(z') - g(z'')}{1 + g(z'')} \right) \right|$$

$$\leq n \frac{\sqrt{2}}{3\sqrt{2}(2+\pi)n/4\pi} + \frac{\pi}{2} \left| \frac{g(z') - g(z'')}{1 + g(z'')} \right|$$

$$< \frac{4\pi}{3(2+\pi)} + \frac{\pi}{2} \frac{4\pi}{3(2+\pi)}$$

$$= \frac{2\pi}{3}.$$

If $z' \in \mathbf{C}_{-1}$ and $z'' \in \mathbf{C}_{-1}$ then

$$\left| \arg \frac{f_d(z')}{f_d(z'')} \right| = \left| \arg \frac{z'^n}{z''^n} \right|$$

$$\leq n|\arg z' - \arg z''|$$

$$\leq n \frac{\sqrt{2}}{3\sqrt{2}(2+\pi)n/4\pi}$$

$$= \frac{4\pi}{3(2+\pi)} < \frac{2\pi}{3}.$$

If $z' \notin \mathbf{C}_{-1}$ and $z'' \in \mathbf{C}_{-1}$ then

$$f_d(z') = f(z') = z'^n(1 + g(z'))$$

and

$$f_d(z'') = z''^n(1 + 0),$$

thus

$$\left| \frac{g(z') - 0}{1 + 0} \right| = |g(z')|$$

$$< \frac{\max |a_j|}{(r - 1)}$$

$$\leq \frac{4(n - 1)}{5n}$$

$$< \frac{4\pi}{3(2 + \pi)} \frac{n - 1}{n}$$

$$< \frac{4\pi}{3(2 + \pi)}.$$

Similarly we have

$$\left| \arg \frac{f_d(z')}{f_d(z'')} \right| = \left| \arg \frac{f(z')}{z''^n} \right| \leq n \left| \arg \frac{z'}{z''} \right| + \left| \arg \left(1 + \frac{g(z') - 0}{1 + 0} \right) \right| < \frac{2\pi}{3}.$$

If $z' \in \mathbf{C}_{-1}$ and $z'' \notin \mathbf{C}_{-1}$ then

$$\left| \arg \frac{f_d(z')}{f_d(z'')} \right| = \left| -\arg \frac{f_d(z'')}{f_d(z')} \right| = \left| \arg \frac{f_d(z'')}{f_d(z')} \right| < \frac{2\pi}{3}.$$

The Lemma is true. ¶

Lemma 2.2   Let $d \geq 0$. Then the all tetrahedra over $\mathbf{C}_d$ passed by the computations lie in n vertical circular cylinders of equal radii $(1 + 0.75n)2^{0.5-d}$ with the centers at the n zeros of f.

Proof.   Since every tetrahedron passed by the computations has exactly one pair of complete triangles, with Theorem 1.2 we know that the distance between some of its vertices and some zeros of the polynomial is not greater than $3n\sqrt{2} 2^{-d}/4 = 0.75n2^{0.5-d}$. Thus the distance between any of its vertices and the zero is not greater than

$$0.75n2^{0.5-d} + \sqrt{2} \cdot 2^{-d} = (1 + 0.75n)2^{0.5-d}.$$

The proof is thus completed.

Notice that the distances concerned are projective distances. ¶

The next lemma is quite technical.

Lemma 2.3 *Let B_d be a circular cylinder with axis $\{0\} \times [d, d+1]$. Let Σ_d be an elementary cube of the triangulation $\mathbf{T}(0; 1)$ between $\mathbf{C}_d$ and $\mathbf{C}_{d+1}$ shown in Figs. 1.3 or 1.4. Let σ_d denote the number of tetrahedra in $\Sigma_d \cap B_d$. Then*

$$\sigma_d \leq \begin{cases} 5 & \mathrm{vol}(\Sigma_d \cap B_d), & \text{if } d = -1, \\ 14 & \mathrm{vol}(\Sigma_d \cap B_d)2^{2d}, & \text{if } d \geq 0, \end{cases}$$

where $\mathrm{vol}(A)$ stands for the volume of A in the Euclidean 3-space.

Proof. For convenience, set $v_d = \mathrm{vol}(\Sigma_d \cap B_d)$. If $d = -1$ then $0 \leq v_d \leq 1$. Refer to Fig. 1.3, it is obvious that $\sigma_d = 0$ when $v_d \leq 1/2$; when $v_d = 1$, $\sigma_d = 5$ since $\Sigma_d \subset B_d$; finally, for the case $1/2 < v_d < 1$, we have $\sigma_d \leq 1$ since every vertical edge of Σ_d intersects four tetrahedra of Σ_d. Hence the lemma is true when $d = -1$.

If $d \geq 0$ then $0 \leq v_d \leq 2^{-2d}$. Referring to Fig. 2.1 (a projection of Fig. 1.4), we discuss the following eight of only possible cases. It is easy to obtain corresponding results for σ_d by simple volume analysis:

$$1^0 \qquad 0 \leq v_d \leq \frac{1}{8}2^{-2d}, \qquad \sigma_d = 0;$$

$$2^0 \qquad \frac{1}{8}2^{-2d} < v_d \leq \frac{1}{4}2^{-2d}, \qquad \sigma_d \leq 1;$$

$$3^0 \qquad \frac{1}{4}2^{-2d} < v_d \leq \frac{3}{8}2^{-2d}, \qquad \sigma_d \leq 2;$$

$$4^0 \qquad \frac{3}{8}2^{-2d} < v_d \leq \frac{1}{2}2^{-2d}, \qquad \sigma_d \leq 4;$$

$$5^0 \qquad \frac{1}{2}2^{-2d} < v_d \leq \frac{3}{4}2^{-2d}, \qquad \sigma_d \leq 7;$$

$$6^0 \qquad \frac{3}{4}2^{-2d} < v_d \leq \frac{7}{8}2^{-2d}, \qquad \sigma_d \leq 8;$$

$$7^0 \qquad \frac{7}{8}2^{-2d} < v_d < 2^{-2d}, \qquad \sigma_d \leq 9;$$

$$8^0 \qquad v_d = 2^{-2d}, \qquad \sigma_d = 14.$$

For example, in case 7^0, a vertex of the square in Fig. 2.1 lies outside B_d, hence at least 5 tetrahedra of Σ_d are not wholly contained in B_d. So, $\sigma_d \leq 9$.

In all the cases, $\sigma_d \leq 14v_d/2^{-2d} = 14v_d 2^{2d}$. This completes the proof.

¶

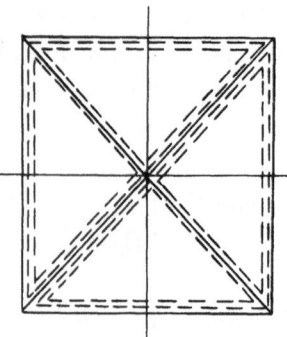

Figure 2.1

The main result of this section is

Theorem 2.4 *For $\epsilon > 0$ small enough, the number of f-evaluations needed for computing the all zeros of the polynomial to an accuracy of ϵ is at most*

$$\pi[5M^2 + 28n(1 + 0.75n)^2 \lceil \log_2(\sqrt{2}(1 + 0.75n)/\epsilon) \rceil],$$

where

$$M = \sqrt{2} + \max\left\{3\sqrt{2}(2 + \pi)n/4\pi, 1 + \frac{5}{4}\frac{n}{n - 1}\max|a_j|\right\}$$

is defined in Lemma 2.1.

Proof. By Lemmas 2.1 and 2.2, all tetrahedra encountered in the computations lie in the base circular cylinder with radius M and axis $\{0\} \times [-1, 0]$ and the n stepped circular cylinders of same radii $(1 + 0.75n)2^{0.5-d}$ between $\mathbf{C}_d$ and $\mathbf{C}_{d+1}$ with axes $\{\xi_j\} \times [0, +\infty)$, $j = 1, \cdots, n$. Refer to Figs. 2.2 and 2.3.

Set

$$D = \lceil \log_2(\sqrt{2}(1 + 0.75n)/\epsilon) \rceil.$$

Notice that the n computational sequences starting from the n (1,2)-labelled edges on ∂Q_m converge respectively to the n zeros of the polynomial, counting multiplicities. Theorem 1.4.1 shows that every computational sequence has exactly one accumulation point. Therefore, when $\epsilon > 0$ is small enough and D big enough, the stepped cylinders with different zeros of the polynomial as centers will certainly separate from each other, and the number of the computational sequences running through $\mathbf{C}_D$ in every stepped cylinder is just equal to the multiplicity of the zero. Notice that, in the view of a dynamic geometry shown in Fig. 2.3, the condition that $\epsilon > 0$ is small enough excludes the possibility of an early truncation of the computations when some computational sequence is still visiting the stepped cylinder of some other zeros. Otherwise, the numerical zeros found may have no correct multiplicities. ·

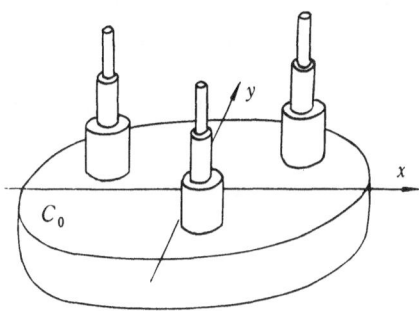

Figure 2.2

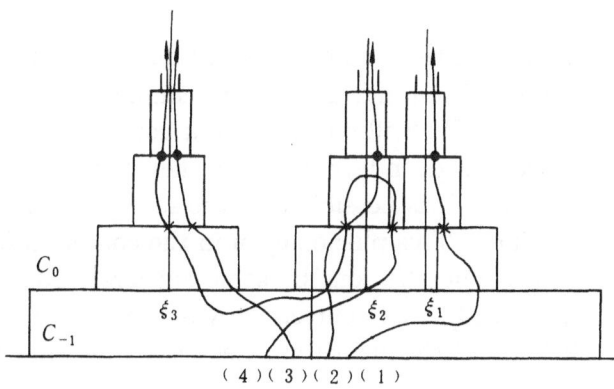

Figure 2.3

According to Lemma 1.4, the computations over $\mathbf{C}_D$ are all unnecessary. Hence, for computing all the zeros of the polynomial to an accuracy of ϵ, the encountered tetrahedra lie in the domain consisted of the base circular cylinder with radius M and the n stepped circular cylinders of height D. By Lemma 2.3, we obtain that the number of tetrahedra within this domain is at most

$$\pi[5M^2 + 28n(1 + 0.75n)^2 \lceil \log_2(\sqrt{2}(1 + 0.75n)/\epsilon)\rceil].$$

According to the algorithm, the implementations of Steps 1 and 2 of the algorithm need no f-evaluations while in Step 3 a pivot from one tetrahedron to the next requires at most one f-evaluation. Thus the total number of evaluations of f is not greater than that of the tetrahedra encountered. This completes the proof. ¶

As the cost for computing one zero of the polynomial is concerned, we have

Corollary 2.5   *The average number of evaluations of f for computing one zero of f within an accuracy of ϵ is not greater than*

$$\pi[5M^2/n + 28(1+0.75n)^2\lceil\log_2(\sqrt{2}(1+0.75n)/\epsilon)\rceil],$$

where M, ϵ are shown in Theorem 2.4. ¶

§3.   Monotonicity Problem

The monotonicity of fixed point algorithms is an important factor which heavily influences the efficiency of the algorithms. Recently, it has received many attentions. Kuhn's algorithm is an example of fixed point algorithms. In the rest of this chapter, we will mainly discuss the monotonicity of Kuhn's algorithm and apply it to compute the all zeros of a polynomial of degree n. We will see that the n computational sequences eventually rise above any given level $\mathbf{C}_d$ until they satisfy the criterion of computational accuracy. It is possible that the sequences may go up and down in turn sometimes. This phenomenon in the computations is called yo-yo behavior of the computations by some authors ([Todd, 1976] and [Allgower & Georg, 1980]). Of course, the yo-yo behavior is a negative factor for the efficiency of the algorithm. Let's see an example first.

Example 3.1(*An example of computations yo-yoing forever*)   We will use Kuhn's algorithm to compute the zeros of the polynomial $f(z) = z^7$. Let the triangulation be $\mathbf{T}(0; 1)$. Since $n = 7$, we can take $m = 5$. A part of vertices of square Q_5 are labelled as in Fig. 2.4. Consider the computational sequence that starts from the first (1,2)-labelled edge on ∂Q_5. For the convenience of showing the computational sequence in the 3-space on the plane, we denote the k-th element (vertex) of the sequence by $(l)_d^k$, where k is the order of the vertex in the sequence, d indicates that the vertex lies in $\mathbf{C}_d$, and l is the label of the vertex. The vertices shown in Fig. 2.4 and their corresponding notations are as follows:

$$(2)^1_{-1} \qquad z = 4 + i,$$
$$(1)^2_{-1} \qquad z = 4,$$
$$(1)^3_{-1} \qquad z = 3,$$
$$(2)^4_{-1} \qquad z = 3 + i,$$
$$(3)^5_{-1} \qquad z = 2 + i.$$

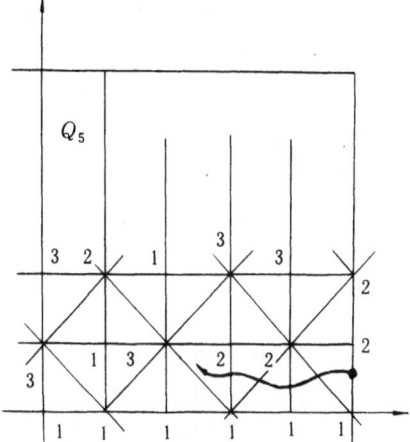

Figure 2.4

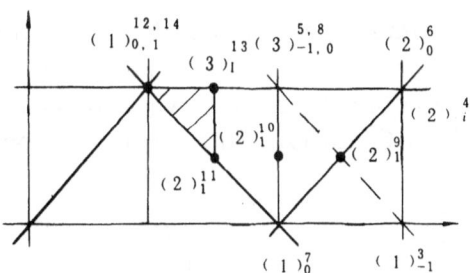

Figure 2.5

After the above four pivots, a complete triangle in $\mathbf{C}_{-1}$ is found. Notice that in this example, all vertices in $\mathbf{C}_{-1}$ and $\mathbf{C}_d (d \geq 0)$ are labelled by z^7. Then referring to Fig. 1.3, with three more 3-dimensional pivots, the next three elements $(2)^6_0, (1)^7_0$ and $(3)^8_0$ are obtained and a complete positive triangle in $\mathbf{C}_0$ is located. Notice that the dotted line in Fig. 2.5 represents an edge in $\mathbf{C}_{-1}$ and $(3)^{5,8}_{-1,0}$ is a short notation for $(3)^5_{-1}$ and $(3)^8_0$. Now refer to Fig. 1.4, we obtain

$$(2)^9_1, (2)^{10}_1, (2)^{11}_1, (1)^{12}_0, (3)^{13}_1, (1)^{14}_1.$$

A positive complete triangle in C_1 is found. The next two vertices $(3)_2^{15}$ and $(3)_1^{16}$ are produced and then a negative complete triangle in C_1 is encountered. Then

$$(1)_0^{17}, (1)_0^{18}, (1)_1^{19}, (2)_2^{20}, (2)_2^{21}, (2)_2^{22}, (1)_1^{23}, (3)_2^{24}, (1)_2^{25}$$

are produced. Now a positive complete triangle in C_2 is obtained.

By the symmetries of both $f(z) = z^7$ and of the triangulation, the next computation will be proceeded according the following rule: Starting from the known positive complete triangle in C_d, after 11 pivots the computation arrives at a positive complete triangle in C_{d+1}. As shown in Fig. 2.6, such complete triangles are more and more approaching the origin as d increases, and finally gives the first numerical zero $z = 0$ of z^7.

Following the sequence from the first vertex to the 25th or the 36th vertex, a clear picture of the geometric structure of the algorithm will be obtained.

Now, let us analyze closely the computation from the positive complete triangle $\{(1)_1^{14}, (2)_1^{11}, (3)_1^{13}\}$ in C_1 to the positive complete triangle $\{(1)_2^{25}, (2)_2^{22}, (3)_2^{24}\}$ in C_2. It is clear that the produced vertices

$$(3)_2^{15}, (3)_1^{16}, (1)_0^{17}, (1)_0^{18}, (1)_1^{19}, (2)_2^{20}, \{(2)_2^{21}, (2)_2^{22}, (1)_1^{23}(3)_2^{24}, (1)_2^{25}$$

by the computation indicate that the sequence rises up from C_1 to C_2 first, and then down from C_2 to C_0, and finally up from C_0 to C_2. A similar analysis will show that the sequence produced by the computation from C_2 to C_3 will rise first from C_2 to C_3 and then down from C_3 to C_1 and finally up from C_1 to C_3. So the computation will be yo-yoing forever.

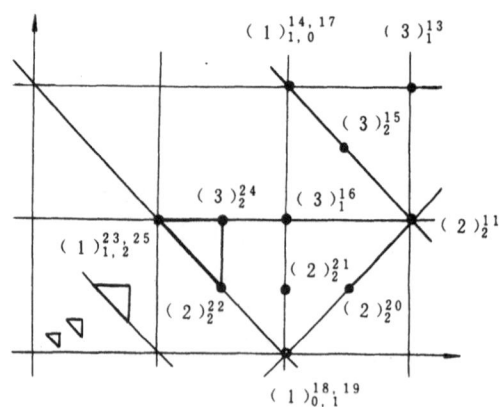

Figure 2.6

Now, we give the definition for monotonicity of the computations of Kuhn's algorithm. Let $\{\sigma_{jk}\}$ be the j-th computational sequence of simplices (Notation 1.3.1). Set $d(\sigma_{jk}) = \max\{d : (z,d) \text{ is a vertex of } \sigma_{jk}\}$.

Definition 3.2   The j-th computational sequence $\{\sigma_{jk}\}$ is said to be monotonously rising if $d(\sigma_{j,k+1}) \geq d(\sigma_{jk})$ for all $k = 1, 2, \cdots$.

Let $\{(z_{jk}, d_{jk})\}$ be the j-th sequence of vertices (Notation 1.3.2).

Definition 3.3   The j-th computational sequence $\{(z_{jk}, d_{jk})\}$ is said to be strongly monotonously rising if $d_{j,k+1} \geq d_{jk}$ for all $k = 1, 2, \cdots$.

It is easy to know that a strongly monotonously rising computational sequence must be monotonously rising. But the inverse is not true.

According to Definitions 3.2 and 3.3, the computational sequence in example 3.1 is neither strongly monotonously rising nor monotonously rising.

As to the monotonicity problem, we also say that a computational sequence is monotonously rising or strongly monotonously rising after some "moment". Next we will prove that every computational sequence which converges to a simple zero of the polynomial must be monotonously rising except at most an early part.

In the rest of this section we use the following assumption.

Assumption 3.4   Assume that $z = \xi_j$ is a simple zero of $f(z)$ for some $1 \leq j \leq n$, and both $\mu = \min_{k \neq j} |\xi_j - \xi_k|$ and $M = \max_{k \neq j} |\xi_j - \xi_k|$ are given. Let $r = \left(M^{n-1} + \dfrac{1}{13} \mu^{n-1} \right)^{1/(n-1)} - M$.

By the assumption, it is clear that $M \geq \mu > 0$ and $r > 0$.

Lemma 3.5   Let $B_j(r) = \{z : |z - \xi_j| < r\}$. Then for any pair of $z', z'' \in B_j(r)$, we have that

$$\left| \arg \left(\frac{f(z')}{f(z'')} \bigg/ \frac{z' - \xi_j}{z'' - \xi_j} \right) \right| < \frac{\pi}{12}$$

and

$$\left| \arg \frac{f(z')}{(z' - \xi_j) \prod_{k \neq j} (\xi_j - \xi_k)} \right| < \frac{\pi}{26}$$

Proof.   Let $\triangle z_j = z - \xi_j$ and $D_{jk} = \xi_j - \xi_k$. Then

$$f(z) = (z - \xi_j) \prod_{k \neq j} (z - \xi_j + \xi_j - \xi_k)$$

$$= \; \Delta z_j \prod_{k \neq j}(\Delta z_j + D_{jk})$$

$$= \; \Delta z_j \Big(\Delta z_j^{n-1} + \Delta z_j^{n-2} \sum_{k \neq j} D_{jk} + \Delta z_j^{n-3} \sum_{\substack{j,k_1,k_2 \\ \text{all different}}} D_{jk_1} D_{jk_2}$$

$$+ \cdots + \Delta z_j \sum_{\substack{j,k_1,\cdots k_{(n-2)} \\ \text{all different}}} D_{jk_1} \cdots D_{jk_{n-2}} + \prod_{k \neq j} D_{jk} \Big)$$

$$\equiv \; \Delta z_j \left[v(\Delta z_j) + \prod_{k \neq j} D_{jk} \right].$$

Obviously, $v(\Delta z_j) \to 0$ as $\Delta z_j \to 0$.

Now, let $z', z'' \in B_j(r)$. Then

$$|v(\Delta z_j')| \; \leq \; |\Delta z_j'|^{n-1} + \binom{n-1}{1}|\Delta z_j'|^{n-2} M + \binom{n-1}{2}|\Delta z_j'|^{n-3} M^2$$

$$+ \cdots + \binom{n-1}{n-2}|\Delta z_j'| M^{n-2}$$

$$= \; (|\Delta z_j'| + M)^{n-1} - M^{n-1}$$

$$< \; \left[\left(M^{n-1} + \frac{1}{13}\mu^{n-1} \right)^{1/(n-1)} - M + M \right]^{n-1} - M^{n-1}$$

$$= \; \mu^{n-1}/13.$$

A similar argument will show that $|v(\Delta z_j'')| < \mu^{n-1}/13$. It is well-known that if $|w| < 1$ then $|\arg(1 + w)| \leq \frac{\pi}{2}|w|$. Let $\Pi = \prod_{k \neq j} D_{jk}$. Then

$$\left| \arg\left(\frac{f(z')}{f(z'')} \Big/ \frac{z' - \xi_j}{z'' - \xi_j} \right) \right|$$

$$= \; \left| \arg \frac{v(\Delta z_j') + \Pi}{v(\Delta z_j'') + \Pi} \right|$$

$$= \; \left| \arg\left(1 + \frac{v(\Delta z_j') - v(\Delta z_j'')}{v(\Delta z_j'') + \Pi} \right) \right|$$

$$\leq \; \frac{\pi}{2} \left| \frac{v(\Delta z_j') - v(\Delta z_j'')}{v(\Delta z_j'') + \Pi} \right|$$

$$< \; \frac{\pi}{2} \frac{\mu^{n-1}/13 + \mu^{n-1}/13}{\mu^{n-1} - \mu^{n-1}/13}$$

$$= \; \frac{\pi}{12}$$

Similarly, we have

$$\left| \arg \frac{f(z')}{(z' - \xi_j) \prod_{k \neq j}(\xi_j - \xi_k)} \right|$$

$$= \left| \arg \frac{v(\triangle z'_j) + \Pi}{\Pi} \right|$$

$$= \left| \arg \left(1 + \frac{v(\triangle z'_j)}{\Pi} \right) \right|$$

$$\leq \frac{\pi}{2} \left| \frac{v(\triangle z'_j)}{\Pi} \right|$$

$$< \frac{\pi}{2} \frac{\mu^{n-1}/13}{\mu^{n-1}}$$

$$= \frac{\pi}{26} . \P$$

According to Lemma 3.5, in calculating the absolute of the argument of $f(z')/f(z'')$ and the absolute of the argument of $f(z')$ for some $z', z'' \in B_j(r)$, the errors caused by substituting the linear function

$$w = (z - \xi_j) \prod_{k \neq j} (\xi_j - \xi_k)$$

for $w = f(z)$ are at most $\pi/12$ and $\pi/26$, respectively.

Lemma 3.6 *Let D be the smallest integer not less than $\log_2(2\sqrt{2}/r)$. Then for any integer $d \geq D$, there is one and only one complete triangle in $B_j(r) \subset \mathbf{C}_d$ and it is also a positive one.*

Proof. Recall the proof of Lemma 1.4.3. In the plane $\mathbf{C}_d$ triangulated into $\mathbf{T}_d(0;1)$, suppose that τ is one of the triangles in which ξ_j lies. Let $T(\tau)$ be the convex hull of the point set union of all triangles in $\mathbf{C}_d$ which share some vertices with τ. Then the distance between any points in $T(\tau)$ and ξ_j is not greater than $2 \cdot \sqrt{2} \cdot 2^{-D} < r$. Hence $T(\tau) \subset B_j(r) \subset \mathbf{C}_d$. As the proof of Lemma 1.4.3, for $d \geq D$, there is a unique complete triangle in $B_j(r) \subset \mathbf{C}_d$ and it is also positive.¶

With the previous lemma, we have

Theorem 3.7 *After crossing the plane $\mathbf{C}_D$, the j-th computational sequence which converges to the simple zero ξ_j is monotonously rising.*

Proof. Consider the cylinder $B_j(r) \times [D, +\infty)$. Obviously, $r < \mu/2$, and there is one and only one complete triangle being positive in $B_j(r) \times$

$\{d\}$ for $d \geq D$. Therefore, the computational sequence converging to ξ_j must enter the cylinder from the complete triangle in $B_j(r) \times \{D\}$, and neither can it pass through the side wall of $B_j(r) \times [D, +\infty)$ and nor return to $B_j(r) \times \{D\}$. So the only possibility is that the sequence rises to cross $B_j(r) \times \{D+1\}$. The sequence will never return to $B_j(r) \times \{D+1\}$, after that, thus it must arrive at $B_j(r) \times \{D+2\}$, and so on. This proves that the sequence is monotonously rising once it passes through $\mathbf{C}_D$. ¶

The following theorem will show that the monotonicity has a great influence on the efficiency of the algorithm.

Theorem 3.8   *Since the plane $\mathbf{C}_D$, with at most five more pivots, the computational sequence converging to the simple zero ξ_j will get on a higher level.*

Proof.   Suppose that the computational sequence converging to ξ_j has arrived at a positive triangle in $\mathbf{T}_d(0; 1), d \geq D$. Without loss of generality, let it be the triangle labelled 1,2 and 3 in Fig. 2.7. We will calculate the label of the point A.

If $l(A) = 2$, we shall consider the vertex B. By Lemma 3.6, $l(B) \neq 2$. By the symmetry, we may assume $l(B) = 1$. Then the triangle $\{B_1, A_2, 3\}$ determines the next vertex C, where the footnote of a vertex indicates its label.

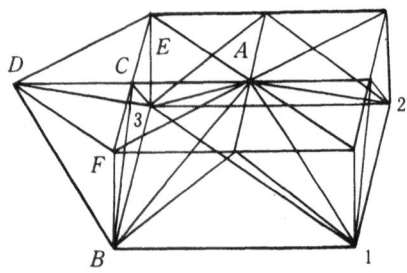

Figure 2.7

If $l(C) = 1$, then $\{C_1, A_2, 3\}$ determines the next vertex E. But $l(E) = 3$, the computation gets in $\mathbf{C}_{d+1}$ by $\{C_1, A_2, E_3\}$. From $\mathbf{C}_d$ to $\mathbf{C}_{d+1}$, there are four pivots.

If $l(C) = 3$, the computation gets in $\mathbf{C}_{d+1}$ by $\{F_1, A_2, C_3\}$. From $\mathbf{C}_d$ to $\mathbf{C}_{d+1}$, there are still four pivots.

If $l(C) = 2$ then $\{B_1, C_2, 3\}$ determines the next vertex D and $l(D) \neq 2$. Otherwise, $\{F_1, E_3, D_2\}$ is a negative complete triangle and it contradicts to Lemma 3.6 (cf. also to the remarks after Lemma 1.4.3). If $l(D) = 1$ or $l(D) = 3$ then the computation gets in $\mathbf{C}_{d+1}$ by $\{D_1, C_2, E_3\}$ or by $\{F_1, C_2, D_3\}$ respectively. In both cases there are five pivots from $\mathbf{C}_d$ to $\mathbf{C}_{d+1}$.

So, if $l(A) = 2$, with at most five pivots, the computational sequence converging to ξ_j must get on a higher level.

The proofs for the cases $l(A) = 1$ or $l(A) = 3$ are similar and then omitted. ¶

The previous theorem can be restated as follows for the computational sequence $\{(z_{jk}, d_{jk})\}$.

Theorem 3.8'   *Under Assumption 3.4, $d_{jk} \geq D + 1$ implies*

$$d_{j,k+5} \geq d_{jk} + 1. ¶$$

In the proof of Theorem 2.4, we utilize such a fact that since $\mathbf{C}_0$, averagely with at most $28\pi(1 + 0.75n)^2$ evaluations of the polynomial, the computational sequence will get on a higher level. But Theorem 3.8 guarantees that the sequence gets on a higher level with at most five evaluations of the polynomial if the computational sequence is monotonously rising. This shows clearly the significance of the monotonicity.

Furthermore, we obtain

Theorem 3.9   *Under Assumption 3.4, there is a positive constant c_j such that $|f(z_{jk})| \leq c_j 2^{-k/5}$ for $k = 1, 2, \cdots$.*

Proof.   Suppose that z_{jk} is the last vertex in $\mathbf{C}_D$ of the computational sequence $\{z_{jk}\}$ which converges to the simple zero ξ_j. By Theorem 3.8, $k - K \leq 5(d - D)$ when $k \geq K$, hence $d \geq k/5 + (D - K/5)$.

On the other hand,

$$f(z) = f'(\xi_j)(z - \xi_j) + \sum_{l=2}^{n} \frac{f^{(l)}(\xi_j)}{l!}(z - \xi_j)^l,$$

so for $k = 1, 2, \cdots$, we have

$$|f(z_{jk})| = \left| f'(\xi_j)(z_{jk} - \xi_j) + \sum_{l=2}^{n} \frac{f^{(l)}(\xi_j)}{l!}(z_{jk} - \xi_j)^l \right|$$

$$= \left| f'(\xi_j)(z_{jk} - \xi_j) \left[1 + \sum_{l=2}^{n} \frac{f^{(l)}(\xi_j)}{l!} (z_{jk} - \xi_j)^{l-1} / f'(\xi_j) \right] \right|$$

$$\leq C_j' |z_{jk} - \xi_j| \leq C_j' (1 + 0.75n) 2^{0.5-d}$$

$$\leq \frac{C_j'(1 + 0.75n)\sqrt{2}}{2^{k/5 + (D-K/5)}}$$

$$= C_j 2^{-k/5},$$

where

$$C_j' = \max_{z \in \sigma_j(r)} \left| f'(\xi_j) \left[1 + \sum_{l=2}^{n} \frac{f^{(l)}(\xi_j)}{l!} (z - \xi_j)^{l-1} / f'(\xi_j) \right] \right|$$

and

$$C_j = \frac{C_j'(1 + 0.75n)\sqrt{2}}{2^{(D-K/5)}}$$

are all positive constants. ¶

§4. Results on Monotonicity

In the previous section we present the problem of monotonicity and obtain that the computational sequence which converges to a simple zero of the polynomial is monotonously rising except at its early part. In this section we continue to discuss the problem.

Lemma 4.1 *If $|z| > \max|a_j| + 1$ then $f(z) \neq 0$. In other words, all the zeros of $f(z)$ lie in $\{z : |z| \leq \max|a_j| + 1\}$.*

Proof. When $|z| > \max|a_j| + 1$ we have

$$|f(z)| = \left| z^n \left(1 + \sum_{j=1}^{n} a_j/z^j \right) \right|$$

$$\geq |z|^n \left(1 - \sum_{j=1}^{n} |a_j|/|z|^j \right)$$

$$\geq |z|^n \left(1 - \max|a_j| \sum_{j=1}^{n} |z|^{-j} \right)$$

$$= |z|^n (1 - \max|a_j|/(|z| - 1)) > 0.$$

The proof is complete. ¶

Now, let

$$\varphi(z) = \sum_{l=0}^{n} |a_l| z^l,$$

$$R = \max |a_j| + 1$$

and

$$H = 1 + \sum_{l=1}^{n} \frac{\varphi^{(l)}(R)}{(l-1)!}.$$

Then for $s = 1, 2, \cdots, n$, we have

$$
\begin{aligned}
|f^{(s)}(\xi_j)| &= \left| \sum_{l=s}^{n} l(l-1) \cdots (l-s+1) a_l \xi_j^{l-s} \right| \\
&\leq \sum_{l=s}^{n} l(l-1) \cdots (l-s+1) |a_l| R^{l-s} = \varphi^{(s)}(R).
\end{aligned}
$$

Lemma 4.2 *Suppose that all the zeros of the polynomial are simple and $0 < \rho \leq \min |f'(\xi_j)|$. If the distance between any vertex of some triangles in $\mathbf{T}_d(\tilde{z}; h)$ and some zero ξ_j of $f(z)$ is not greater than $\sigma = \min\{1, \rho/5H\}$, then the triangle is not a negative complete triangle.*

Proof. Since $f(\xi_j) = 0$, by Taylor's formula,

$$f(z) = f'(\xi_j)(z - \xi_j) + \sum_{l=2}^{n} \frac{f^{(l)}(\xi_j)}{l!}(z - \xi_j)^l.$$

Denote by $\{\alpha, \beta, \gamma\}$ the triangle concerned, and assume that the two right-sides of the triangle intersect at γ. Then we have

$$
\begin{aligned}
&\frac{f(\beta) - f(\gamma)}{f(\alpha) - f(\gamma)} \\
&= \frac{f'(\xi_j)(\beta - \gamma)) + \sum_{l=2}^{n}(f^{(l)}(\xi_j)/l!)[(\beta - \xi_j)^l - (\gamma - \xi_j)^l]}{f'(\xi_j)(\alpha - \gamma)) + \sum_{l=2}^{n}(f^{(l)}(\xi_j)/l!)[(\alpha - \xi_j)^l - (\gamma - \xi_j)^l]} \\
&= \frac{\beta - \gamma}{\alpha - \gamma}\left[1 + \left\{ \left[\sum_{l=2}^{n}(f^{(l)}(\xi_j)/l!) \sum_{s=1}^{l}(\beta - \xi_j)^{l-s}(\gamma - \xi_j)^{s-1} \right. \right. \right. \\
&\quad \left. - \sum_{l=2}^{n}(f^{(l)}(\xi_j)/l!) \sum_{s=1}^{l}(\alpha - \xi_j)^{l-s}(\gamma - \xi_j)^{s-1} \right] \\
&\quad \left. \left/ \left[f'(\xi_j) + \sum_{l=2}^{n}(f^{(l)}(\xi_j)/l!) \sum_{s=1}^{l}(\alpha - \xi_j)^{l-s}(\gamma - \xi_j)^{l-1} \right] \right\} \right] \\
&\equiv \frac{\beta - \gamma}{\alpha - \gamma}(1 + \Theta).
\end{aligned}
$$

But the distances between α, β, γ and ξ_j are not greater than σ, hence

$$
\begin{aligned}
|\Theta| &\leq \frac{2\sum_{l=2}^{n}(\varphi^{(l)}(R)/l!)l\sigma^{l-1}}{|f'(\xi_j)| - \sum_{l=2}^{n}(\varphi^{(l)}(R)/l!)l\sigma^{l-1}} \\
&\leq \frac{2\sigma\sum_{l=2}^{n}\varphi^{(l)}(R)/(l-1)!}{\rho - \sigma\sum_{l=2}^{n}\varphi^{(l)}(R)/(l-1)!} \\
&= \frac{2\sigma H}{\rho - \sigma H} \\
&\leq 12.
\end{aligned}
$$

We know that for $|w| \leq 1/2$,

$$
\arg(1+w)| \leq \arcsin|w| \leq \arcsin\frac{1}{2} = \frac{\pi}{6},
$$

thus we have

$$
\begin{aligned}
\frac{\pi}{3} &= \frac{\pi}{2} - \frac{\pi}{6} = \arg\frac{\beta-\gamma}{\alpha-\gamma} - \frac{\pi}{6} \leq \arg\frac{f(\beta)-f(\gamma)}{f(\alpha)-f(\gamma)} \\
&\leq \arg\frac{\beta-\gamma}{\alpha-\gamma} + \frac{\pi}{6} = \frac{\pi}{2} + \frac{\pi}{6} \\
&= \frac{2\pi}{3}.
\end{aligned}
$$

Suppose that $\{\alpha, \beta, \gamma\}$ is a negative complete triangle and the labels of α, β, γ are 1,3,2, respectively. In this case, if $f(\alpha), f(\beta)$ and $f(\gamma)$ lie on a same line, then

$$
\arg\frac{f(\beta)-f(\gamma)}{f(\alpha)-f(\gamma)} = \pi \text{ or } 0.
$$

If the triangle $\{f(\alpha), f(\beta), f(\gamma)\}$ contains the origin shown in Fig. 2.8(1), then

$$
\arg\frac{f(\beta)-f(\gamma)}{f(\alpha)-f(\gamma)} < 0.
$$

If $f(\gamma)$ and the origin lie in different sides of the line AB shown in Fig. 2.8(2) and (3), where $A = f(\alpha), B = f(\beta)$, then

$$
\arg\frac{f(\beta)-f(\gamma)}{f(\alpha)-f(\gamma)} < \pi/3.
$$

If $f(\gamma)$ and the origin lie in same side of the line AB shown in Fig. 2.8(4) and (5), then

$$
\arg\frac{f(\beta)-f(\gamma)}{f(\alpha)-f(\gamma)} > 2\pi/3 \text{ or } < 0.
$$

All these cases will lend to contradictions. Therefore, $\{\alpha, \beta, \gamma\}$ is not a negative complete triangle. ¶

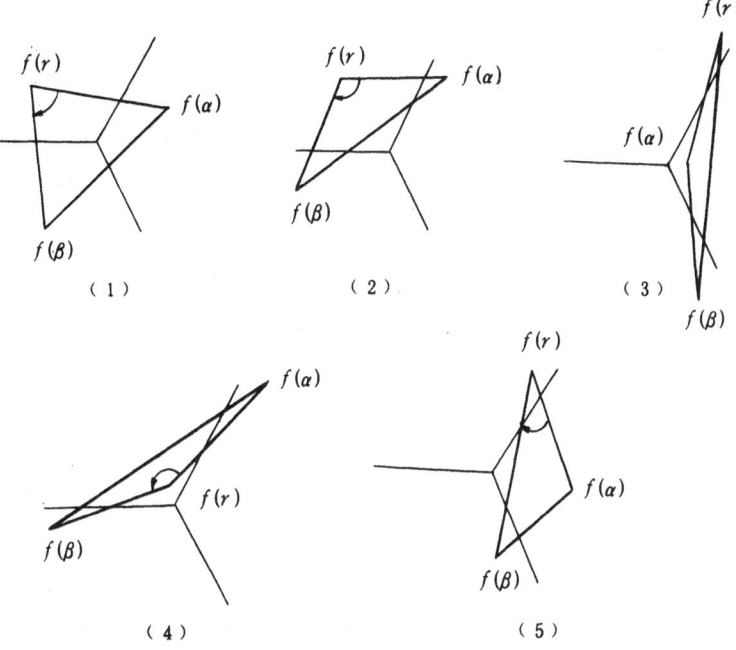

Figure 2.8

Lemma 4.3 *Suppose that all the zeros of $f(z)$ are simple. Let $D = \lceil \log_2[(1 + 0.75n)\sqrt{2}/\sigma] \rceil$, here $\sigma = \min\{1, \rho/5H\}$ defined in Lemma 4.2. Then for $d \geq D$, there exists no negative complete triangles in $\mathbf{T}_d(\tilde{z}; 1)$.*

Proof. Suppose otherwise that $\{\alpha, \beta, \gamma\}$ is such a triangle. Thus Lemma 2.2 implies that the distances between α, β, γ and some zeros of $f(z)$ are all not greater than $(1 + 0.75n)2^{0.5-d} \leq (1 + 0.75n)2^{0.5-D} \leq \sigma$. This contradicts Lemma 4.2. ¶

The following theorem is directly from Lemma 4.3.

Theorem 4.4 *Under the conditions of Lemma 4.3, if $d(\sigma_{jk}) \geq D$ then $d(\sigma_{j,k+1}) \geq d(\sigma_{jk})$ for $j = 1, \cdots, n$. That is, every computational sequence is monotonously rising since $\mathbf{C}_D$.* ¶

Let $\{\sigma_{jk}\}$ be a computational sequence of simplices. Call d the height of the plane $\mathbf{C}_d$. When $\dim(\sigma_{jk}) = 3$, define $D(\sigma_{jk})$ to be the sum of the heights of the planes in which the four vertices of σ_{jk} lie. We know that $d(\sigma_{jk})$ monotonously rising means the least height of the planes in which some vertex of the simplex lie doesn't decrease while $D(\sigma_{jk})$ monotonously rising means that the height $D(\sigma_{jk})/4$ of the barycenter of the simplex doesn't decrease. It is evident that $D(\sigma_{jk})$ monotonously rising guarantees that $d(\sigma_{jk})$ is monotonously rising.

Theorem 4.5   Under the conditions of Lemma 4.3, if $d(\sigma_{jk}) \geq D$ then $D(\sigma_{j,k+1}) \geq D(\sigma_{jk})$ for $j = 1, \cdots, n$.

Proof.   According to the algorithm, when $D(\sigma_{jk}) = 4d(\sigma_{jk}) + 1$, σ_{jk} is as $\{\alpha, \beta, \gamma, \delta\}$ shown in Fig. 2.9. If $\{\alpha, \beta, \gamma\}$ is its exit then $\{\alpha, \beta, \gamma\}$ is a negative complete triangle in $\mathbf{C}_{d(\sigma_{jk})}$. This contradicts Lemma 4.3, if $\{\alpha, \delta, \gamma\}$ is its exit then $D(\sigma_{j,k+1}) = D(\sigma_{jk})$. If $\{\alpha, \beta, \delta\}$ or $\{\beta, \gamma, \delta\}$ is its exit then $D(\sigma_{j,k+1}) = D(\sigma_{jk}) + 1$.

When $D(\sigma_{jk}) = 4d(\sigma_{jk}) + 2$, σ_{jk} is as $\{\alpha, \beta, \delta, \epsilon\}$ shown in Fig. 2.9. If $\{\alpha, \delta, \beta\}$ is its exit, denote the projection of δ on $\mathbf{C}_{d(\sigma_{jk})}$ by δ', then $\{\alpha, \beta, \delta'\}$ is a negative complete triangle. This also contradicts to Lemma 4.3. If $\{\alpha, \beta, \epsilon\}$ is its exit then $D(\sigma_{j,k+1}) = D(\sigma_{jk})$. If $\{\alpha, \epsilon, \delta\}$ or $\{\beta, \delta, \epsilon\}$ is its exit then $D(\sigma_{j,k+1}) = D(\sigma_{jk}) + 1$.

When $D(\sigma_{jk}) = 4d(\sigma_{jk}) + 3$, then σ_{jk} is as $\{\beta, \delta, \epsilon, \zeta\}$ shown in Fig. 2.9. By the definitions of the triangulation and of the labelling, both $\{\beta, \zeta, \delta\}$ and $\{\beta, \zeta, \epsilon\}$ are not the exit of $\{\beta, \delta, \epsilon, \zeta\}$. If $\{\beta, \epsilon, \delta\}$ is its exit, denote by ϵ' the projection of ϵ on $\mathbf{C}_{d(\sigma_{jk})}$, then $\{\beta, \delta', \epsilon'\}$ is a negative complete triangle. This is in contradiction with Lemma 4.3. So only $\{\zeta, \beta, \epsilon\}$ may be its exit. Thus $D(\sigma_{j,k+1}) = D(\sigma_{jk}) + 2$. The theorem is thus true. ¶

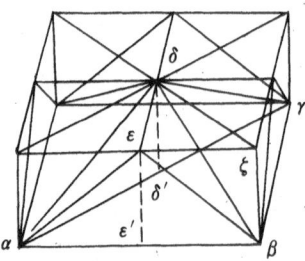

Figure 2.9

· Similarly, by improving Theorem 3.7 we can provide the result in the form of $D(\sigma_{j,k+1}) \geq D(\sigma_{jk})$.

Now, we consider the description of an upper bound of the beginning moment D of the monotonous computations with the coefficients of the polynomial.

We have known that

$$D = \lceil \log_2[(1 + 0.75n)\sqrt{2}/\min\{1, \rho/5H\}] \rceil,$$

where both $H = 1 + \sum_{l=2}^n \varphi^{(l)}(R)/(l-1)!$ and $R = \max|a_j| + 1$ can be described with the coefficients $a_1, \cdots, a_n$. So it suffices to describe ρ with the coefficients.

Since $f(z) = \prod_{l=1}^n (z - \xi_l)$, we have $f'(\xi_j) = \prod_{l \neq j}(\xi_j - \xi_l)$. By the properties of Vandermonde determinant, for any fixed j

$$
\begin{aligned}
\triangle &= \sqrt{\left| \det \begin{pmatrix} n & s_1 & s_2 & \cdots & s_{n-1} \\ s_1 & s_2 & s_3 & \cdots & s_n \\ \cdots & & & & \\ s_{n-1} & s_n & s_{n+1} & \cdots & s_{2n-2} \end{pmatrix} \right|} \\
&= \sqrt{|\det A \det A^T|} \\
&= \prod_{s>l} |\xi_s - \xi_l| \\
&\leq (2R)^{n(n-1)/2-(n-1)} \prod_{l \neq j} |\xi_j - \xi_l|,
\end{aligned}
$$

where $s_k = \sum_{l=1}^n \xi_l^k$ and

$$
A = \begin{pmatrix} 1 & 1 & \cdots & 1 \\ \xi_1 & \xi_2 & \cdots & \xi_n \\ \xi_1^2 & \xi_2^2 & \cdots & \xi_n^2 \\ \cdots & & & \\ \xi_1^{n-1} & \xi_2^{n-1} & \cdots & \xi_n^{n-1} \end{pmatrix}.
$$

Hence $|f'(\xi_j)| = \prod_{l \neq j} |\xi_j - \xi_l| \geq \triangle/(2R)^{n(n-1)/2-(n-1)}$.

Set $\rho = \triangle/(2R)^{n(n-1)/2-(n-1)}$. Since the determinant used in $\triangle$ is a symmetric polynomial in $\xi_1, \cdots, \xi_n$, it can be regarded as a polynomial of the basic symmetric polynomials

$$\sum_{i_1 < \cdots < i_k} \xi_{i_1} \cdots \xi_{i_k} = (-1)^k a_k$$

with the lexicographical system (cf. [Jacobson, 1974]). Therefore, ρ can be described with the coefficients of the polynomial.

Thus an upper bound for the beginning moment is obtained.

Chapter 3

Newton Method and Approximate Zeros

In this chapter, based on [Smale 1981], we discuss some problems in utilizing Newton method to find a zero of a complex polynomial, present certain criteria for the convergence of the method, and provide estimates for its convergent speed. To do these, the so-called approximate zero proposed by S. Smale is a key concept.

In Section 1, we present the concept of approximate zeros, and in Section 2 we discuss the connection between the coefficients of a polynomial and its critical values. Then we provide an estimate of the effect of one step of Newton iteration in Section 3. Finally, we give both the convergent criteria and the convergent speed estimates for Newton method in Section 4.

In Section 2 we will quote some preliminary results of complex analysis and of homotopy theory. The key points are Loewner Theorem and Mapping-lifting Theorem. In addition to indicating their references, we also discuss in details their significances and usages.

§1. Approximate Zeros

Newton method has a history of nearly three hundred years. There is a great deal of literature on it, and people have accumulated rich experiences in practice its usage. Besides, the formulation of this method is very simple. All these often lead people to have a preference to employ it in the problems of zero-finding.

However, contrasting sharply with Kuhn's algorithm described in the previous chapters, Newton method does not guarantee global convergence. Whether it is convergent, is heavily dependent on its starting point, or in other words, on the initial value of the iteration. If the initial value lies in its quick convergent domain, the convergent speed of the method is very high. Otherwise, the convergent speed can be very low, or even the method is not convergent. Thus, the selection of an initial value for the method is always a serious problem, in particular for non-specialists. In this sense, Newton method is in fact not so easy to use in practice.

Consider again the problem of computing the zeros of a monic complex polynomial $f(z) = z^n + a_1 z^{n-1} + \cdots + a_n$ of degree n.

Definition 1.1 (Newton method)   Taking $z_0 \in \mathbf{C}$ as an initial value, define inductively the iterations as

$$z_k = z_{k-1} - f(z_{k-1})/f'(z_{k-1}), \quad k = 1, 2, \cdots.$$

As stated above, the choose of the initial value z_0 is of great importance. If the initial value is selected near some exact zeros then the convergent speed of Newton method is fairly high. Therefore, the most important thing is to find a good initial value, or to determine the quick convergent domain. For this, S. Smale proposed the following concept of approximate zeros.

Definition 1.2 z_0 is called an approximate zero of $f(z)$ provided z_k in Definition 1.1 is well defined for all k, the sequence $\{z_k\}$ converges to z^* as $k \to \infty$ with $f(z^*) = 0$, and $f(z_k)/f(z_{k-1}) < 1/2$ for all $k = 1, 2, \cdots$.

It is important to note that whether a point is an approximate zero or not depends on the polynomial concerned.

The step size of the above-presented Newton method is 1, now we define a kind of Newton methods with step size $h \in (0, 1]$.

Definition 1.3 (Modified Newton method)   Let $z_0 \in \mathbf{C}$ and define inductively $z_k = N_{h,f}(z_{k-1})$, where

$$N_{h,f}(z) = z - hf(z)/f'(z),$$

here h is a chosen number in $(0, 1]$.

If $h = 1$, this is exactly the original Newton method.

§2.   Coefficients of Polynomials

In this section we discuss the connection between the coefficients of a polynomial and its critical points. The main result is Theorem 2.5. The proof of the theorem requires two aspects of preliminary knowledge. One is Loewner Theorem, the other is the results used in the proof of Lemma 2.3. We will point out the reference for Loewner Theorem and provide also some remarks related to Theorems 2.4 and 2.5. To highlight the main idea of this section, the proof of Lemma 2.3 is placed at the end of the section.

As before, let $\mathbf{C}$ be the complex z-plane and $\mathbf{C'}$ be the w-plane. Let $D_r = \{w \in \mathbf{C'} : |w| < r\}$. $\theta \in \mathbf{C}$ is called a critical point of f if $f'(\theta) = 0$, and in this case $f(\theta)$ is called a critical value of f.

Theorem 2.1 (Loewner Theorem)   *Let* $g : D_1 \to \mathbf{C}$ *be a one-to-one analytic function which can be expressed as a convergent power series* $g(w) = \sum_{i=0}^{\infty} b_i w^i$, $|w| < 1$ *with* $b_0 = 0$ *and* $b_1 = 1$. *Suppose that* $f : \Omega \to D_1$ *is an inverse to* g *with* $0 \in \Omega$. *Let* $f(z) = \sum_{i=0}^{\infty} a_i z^i$ *near* $z = 0$. *Then*

$$|a_k| \le B_k, \qquad k = 1, 2, \cdots,$$

where

$$B_k = 2^k \cdot \frac{1 \cdot 3 \cdots (2k-1)}{1 \cdot 2 \cdots (k+1)}. \P$$

For a proof of Loewner Theorem, refer to [Jenkins, 1965]. Notice that under the conditions of the theorem, $a_0 = 0$ and $a_1 = 1$.

We will need a slight extension of this theorem , that is

Theorem 2.2 (Extended Loewner Theorem)   *Let* $g : D_R \to \mathbf{C}$, $R > 0$, *be a one-to-one analytic function which can be expressed as a convergent power series* $g(w) = \sum_{i=0}^{\infty} b_i w^i$, $|w| < R$, *with* $b_0 = 0$ *and* $b_1 \ne 0$. *Let* $f : \Omega \to D_R$ *be an inverse to* g *with* $0 \in \Omega$ *and* $f(z) = \sum_{i=0}^{\infty} a_i z^i$ *near* $z = 0$. *Then for* $k = 1, 2, \cdots$,

$$\left| \frac{a_k}{a_1} \right|^{1/(k-1)} \cdot \frac{R}{|a_1|} \le B_k^{1/(k-1)}.$$

Proof.   First, we discuss the case $R = 1$. Notice that $a_1 = f'(0) = 1/g'(0) = 1/b_1$. Let $g_0(w) = g(w)/b_1$, $f_0(z) = f(z/a_1)$. Then for $|w| < 1$ we have

$$f_0(g_0(w)) = f\left(\frac{g(w)/b_1}{a_1} \right) = f(g(w)) = w.$$

But

$$f_0(z) = z + (a_2/a_1^2)z^2 + (a_3/a_1^3)z^3 + \cdots.$$

Then applying Loewner Theorem to g_0, we have $|a_k/a_1^k| \leq B_k$. So

$$\left|\frac{a_k}{a_1}\right|^{1/(k-1)} \cdot \frac{1}{|a_1|} \leq B_k^{1/(k-1)} \quad \text{for} \quad k = 2, 3, \cdots.$$

Hence, the theorem is true when $R = 1$.

Now, suppose that $R > 0$. Let $g_1(w) = g(Rw)$, $f_1(z) = f(z)/R$. Then for $|Rw| < R$ we have

$$f_1(g_1(w)) = f_1(g(Rw)) = f(g(Rw))/R = Rw/R = w.$$

Notice that $f_1(z) = (a_1/R)z + (a_2/R)z^2 + \cdots$. Applying the result of the case $R = 1$ to g_1 and f_1, we have that for $k = 2, 3, \cdots$,

$$\left|\frac{a_k/R}{a_1/R}\right|^{1/(k-1)} \cdot \frac{1}{|a_1/R|} \leq B_k^{1/(k-1)},$$

that is,

$$\left|\frac{a_k}{a_1}\right|^{1/(k-1)} \frac{R}{|a_1|} \leq B_k^{1/(k-1)}. \P$$

Lemma 2.3 *Let $f(z)$ be a polynomial of degree n with $f(0) = 0$ and $R = \min_{\theta, f'(\theta)=0} |f(\theta)| > 0$. Then there is a one-to-one analytic function $g : D_R \to \mathbf{C}$ such that $g(0) = 0$ and $f(g(w)) = w$ for all $w \in D_R$.* ¶

The proof of this lemma is presented at the end of this section. Now we use it to prove the main result of this section.

Theorem 2.4 *Let $f(z) = a_1 z + a_2 z^2 + \cdots + a_n z^n$ with $a_1 \neq 0$. Then there is a critical point $\theta \in \mathbf{C}$ of f such that for all $k = 2, 3, \cdots, n$,*

$$\left|\frac{a_k}{a_1}\right|^{1/(k-1)} \cdot \left|\frac{f(\theta)}{a_1}\right| \leq \beta_k,$$

where

$$\beta_k = B_k^{1/(k-1)}.$$

Proof. From the assumptions of the theorem, $f(0) = 0$. When $R = \min_{\theta, f'(\theta)=0} |f(\theta)| = 0$, it is clear that the theorem is true. If $R > 0$, Lemma 2.3 gives a one-to-one analytic function $g : D_R \to \mathbf{C}$, $g(w) = \sum_{i=0}^{\infty} b_i w^i$ such that $g(0) = 0$ and $f(g(w)) = w$ for all $w \in D_R$.

Notice that $g(0) = 0$ implies $b_0 = 0$, and $f'(0)g'(0) = 1$ gives $b_1 \neq 0$. Now, the theorem follows directly from the extended Loewner Theorem.

¶

From $\beta_k = B_k^{1/(k-1)}$ we obtain easily that $\beta_2 = 2$, $\beta_3 = \sqrt{5}$, $\beta_4 = (14)^{1/3}$, $\beta_5 = (42)^{1/4} \approx 2.55$, $\beta_6 \approx 2.61$, $\beta_7 \approx 2.65, \cdots, \beta_{20} \approx 3.29, \cdots$, and so on.

Now, we give a uniform estimate for all β_k as follows.

Theorem 2.5 Let $f(z) = a_1 z + a_2 z^2 + \cdots + a_n z^n$ with $a_1 \neq 0$. Then there is a critical point $\theta \in \mathbf{C}$ of f such that for all $k = 2, 3, \cdots$,

$$\left| \frac{a_k}{a_1} \right|^{1/(k-1)} \left| \frac{f(\theta)}{a_1} \right| \leq K,$$

where $K = 4$.

Proof. Since

$$
\begin{aligned}
B_k &= 2^{k-1} \cdot \frac{2}{k+1} \cdot \frac{1 \cdot 3 \cdots (2k-1)}{1 \cdot 2 \cdots k} \\
&\leq 2^{k-1} \frac{2}{k+1} \cdot 2^{k-1} \\
&\leq 4^{k-1},
\end{aligned}
$$

we have $\beta_k \leq 4$. ¶

Remark 2.6 Theorems 2.4 and 2.5 are true only for polynomials. The example $f(z) = (1/\alpha)e^{\alpha z} - 1/\alpha$ with $\alpha > 1$ shows that the theorems are no longer true even for analytic functions. In fact, $f(z)$ in the example has no critical point at all.

Remark 2.7 In Theorem 2.5 we have proved that K could be taken 4. The next example shows that $K \geq 1$.

In Theorem 2.5, K is independent of n and k. For every n and k, denote by $K_{n,k}$ the lower bound of K that makes Theorem 2.5 true for the polynomial

$$f(z) = \begin{cases} (z-1)^n - (-1)^n, & \text{if } k < n, \\ z^n - nz, & \text{if } k = n. \end{cases}$$

When $k = n$, $f'(z) = n(z^{n-1} - 1)$. If $f'(\theta) = 0$ then $\theta^{n-1} = 1$. Thus $f(\theta) = \theta(\theta^{n-1} - n)$ and so $R = \min_{\theta, f'(\theta)=0} |f(\theta)| = n - 1$. Since $a_1 = -n$ and $a_n = 1$, we have

$$K_{n,k} = \left(\frac{1}{n} \right)^{1/(n-1)} \cdot \frac{n-1}{n}.$$

When $k < n$, $f'(z) = n(z-1)^{n-1}$. So $f'(\theta) = 0$ implies $\theta = 1$, and thus $f(\theta) = (-1)^{n+1}$. Hence $R = 1$. Now, $a_1 = (-1)^{n-1}n$ and

$$a_k = (-1)^{n-k}\frac{n!}{k!(n-k)!},$$

and so

$$K_{n,k} = \left(\frac{(n-1)!}{k!(n-k)!}\right)^{1/(k-1)} \cdot \frac{1}{n}.$$

Since

$$\sup_{n>2} K_{n,2} = \sup_{n>2}\frac{(n-1)!}{2(n-2)!} \cdot \frac{1}{n}$$
$$= \sup_{n>2}\frac{n-1}{2n}$$
$$= \frac{1}{2},$$

and

$$K_{2,2} = \frac{1}{4},$$

we have $\sup_n K_{n,2} = 1/2$.

Similarly, since

$$\sup_{n>3} K_{n,3} = \sup_{n>3}\left(\frac{(n-1)!}{6(n-3)!}\right)^{1/2} \cdot \frac{1}{n}$$
$$= \sup_{n>3}\left(\frac{(n-1)(n-2)}{6}\right)^{1/2} \cdot \frac{1}{n}$$
$$= \left(\frac{1}{6}\right)^{1/2}$$

and

$$K_{3,3} = \left(\frac{1}{3}\right)^{1/2} \cdot \frac{2}{3} = \left(\frac{4}{27}\right)^{1/2} < \left(\frac{1}{6}\right)^{1/2},$$

we have $\sup_n K_{n,3} = (1/6)^{1/2}$.

When $k > 3$, because

$$\sup_{n>k} K_{n,k} = \sup_{n>k}\left(\frac{(n-k+1)(n-k+2)\cdots(n-1)}{k!\,n^{k-1}}\right)^{1/(k-1)}$$
$$= \left(\frac{1}{k!}\right)^{1/(k-1)},$$

and

$$\sup_{n=k} K_{n,k} = \left(\frac{1}{k}\right)^{1/(k-1)} \cdot \frac{k-1}{k} > \left(\frac{1}{k!}\right)^{1/(k-1)},$$

we obtain $\sup_n K_{n,k} = (1/k)^{1/(k-1)}((k-1)/k)$ for $k > 3$.

In summary, notice that

$$\lim_{k\to\infty} \sup_n K_{n,k} = \lim_{k\to\infty} \left(\frac{1}{k}\right)^{1/(k-1)} \frac{k-1}{k} = 1$$

and K is independent of n and k, we obtain $K \geq 1$.

Remark 2.8 S. Smale suspects that $K = 1$. This problem has a close relationship with Bieberbach conjecture (cf. Chapter 5).

Conjecture 2.9 (Bieberbach Conjecture) Let $f : D_1 \to \mathbf{C}'$ be a one-to-one analytic function and $f(z) = z + a_2 z^2 + a_3 z^3 + \cdots$. Then $|a_k| \leq k$ for $k = 2, 3, \cdots$.

De Branges proved in 1985 that Bieberbach conjecture is true. This is a great achievement in the world of both pure and applied mathematics.

Remark 2.10 For the possible improvement of K, in the next discussions we use Theorem 2.5 in the form

$$\left|\frac{a_k}{a_1}\right|^{1/(k-1)} \left|\frac{f(\theta)}{a_1}\right| \leq K.$$

Remember that $K \geq 1$, and we can take always $K = 4$.

Next, we will present a proof of Lemma 2.3. The main tool used is the Mapping-lifting Theorem. We first give the concept of covering spaces.

Definition 2.11 Let the topological spaces X and $\hat{X}$ be connected and locally path-connected, and $p : \hat{X} \to X$ be a continuous function. $(\hat{X}, p)$ is called a covering space of X if for every $x \in X$, there exists a connected open set U containing x such that

(1) $p^{-1}(U) = \bigcup_{\alpha \in \Lambda_x} S_\alpha$, where all S_α are open sets in $\hat{X}$ which don't intersect each other; and

(2) for every $\alpha \in \Lambda_x$, $p|S_\alpha : S_\alpha \to U$, the restriction of p on S_α, is a homomorphism.

Theorem 2.12 (Mapping-lifting Theorem) *Let $(\hat{X}, p)$ be a covering space of X, $x_0 \in X$, $\hat{x}_0 \in p^{-1}(x_0)$ and Y be a connected and locally path-connected space. Then for any continuous mapping $\varphi : (Y, y_0) \to (X, x_0)$,*

$\varphi_*(\pi_1(Y, y_0)) \subset p_*(\pi_1(\hat{X}, \hat{x}_0))$ *if and only if there exists a unique contin-*
uous mapping $\varphi' : (Y, y_0) \to (\hat{X}, \hat{x}_0)$ *such that* $p \circ \varphi' = \varphi$, *where* $p \circ \varphi'$ *is*
defined by $(p \circ \varphi')(y) = p(\varphi'(y))$ *for all* $y \in Y$.

See [Christenson & Voxman, 1977] for a proof of the theorem. In
the theorem, $\varphi : (Y, y_0) \to (X, x_0)$ means $\varphi : Y \to X$ and $\varphi(y_0) = x_0$. $\pi_1(Y, y_0)$ is the fundamental group of Y with the base point y_0, and
the meanings of $\pi_1(\hat{X}, \hat{x}_0)$ and $\pi_1(X, x_0)$ are similar. Now we have the
continuous mappings $p : (\hat{X}, \hat{x}_0) \to (X, x_0)$ and $\varphi : (Y, y_0) \to (X, x_0)$,
they induce respectively the homomorphisms $p_* : \pi_1(\hat{X}, \hat{x}_0) \to \pi_1(X, x_0)$
and $\varphi_* : \pi_1(Y, y_0) \to \pi_1(X, x_0)$. Mapping-lifting Theorem shows that,
there is a unique continuous mapping $\varphi' : (Y, y_0) \to (\hat{X}, \hat{x}_0)$ satisfying
$p \circ \varphi' = \varphi$ if and only if $\varphi_*(\pi_1(Y, y_0)) \subset p_*(\pi_1(\hat{X}, \hat{x}_0))$. In this case,
$\varphi' : (Y, y_0) \to (\hat{X}, \hat{x}_0)$ is called a lifting of $\varphi : (Y, y_0) \to (X, x_0)$ to the
covering space. Under the conditions of the theorem, the lifting is unique.

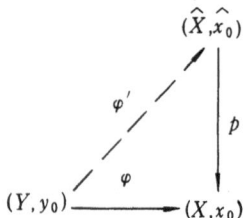

It is well-known that the fundamental group of a contractible space
is trivial. In other words, if X can be contracted into one point x_0 then
$\pi_1(X, x_0) = \{e\}$, a group with only one element. Of course, the fundamental group of a disk with any point in it as a base point is trivial.

The other tool used in the proof of Lemma 2.3 is Inverse Function
Theorem. Let $U \subset \mathbf{R}^m$ be an open set and $f : U \to \mathbf{R}^m$ be a mapping. f
is said to be a C^r mapping if all of its k-th partial derivatives exist and are
continuous on U for $k = 1, 2, \cdots, r$. Similarly, f is called a C^∞ mapping if
for any $k = 1, 2, \cdots$, all its k-th partial derivatives exist and are continuous
on U. If f is real analytic then f is said to be a C^ω mapping. We have

Theorem 2.13 (Inverse Function Theorem)   *Let* $U \subset \mathbf{R}^n$ *be an open*
set and $f : U \to \mathbf{R}^n$ *be a* C^r *mapping, where* $r = 1, 2, \cdots$, *or* ∞, *or* ω.
If $x \in U$ *and the Jacobian matrix* Df_x *of* f *at* x *is invertible then* f *is a*
local C^r-*diffeomorphism at* x_0.

Now we are ready to prove Lemma 2.3.

The proof of Lemma 2.3. Let

$$U = \mathbf{C} - \bigcup_{\theta, f'(\theta)=0} f^{-1}(f(\theta))$$

and

$$V = \mathbf{C}' - \bigcup_{\theta, f'(\theta)=0} \{f(\theta)\}.$$

It is obvious that both U and V are connected and locally path-connected space. Consider $f : U \to V$. Of course, f is continuous.

For $w \in V$, consider the polynomial $F : \mathbf{C} \to \mathbf{C}'$ defined by $F(z) = f(z) - w$. By Fundamental Theorem of Algebra, there are $x_1, \cdots, x_n \in \mathbf{C}$ such that $F(x_j) = 0$, that is, $f(x_j) = w$, and so $F'(x_j) = f'(x_j) \neq 0$ for $j = 1, \cdots, n$. Hence $x_j \neq x_i$ for $1 \leq i < j \leq n$ and $x_j \in U$ for $j = 1, \cdots, n$.

Since for every $j = 1, \cdots, n$, $f'(x_j) \neq 0$, by Inverse Function Theorem, f is a local diffeomorphism at x_j, i.e., there exists an open set $S_j^* \subset U$ such that $x_j \in S_j^*$ and the restriction $f|S_j^* : S_j^* \to f(S_j^*)$ is a diffeomorphism. Without loss of generality, suppose that all S_j^* are bounded and don't intersect each other. Because $f(z)$ is a polynomial, there is a positive number M such that $|f(z)| \geq 1$ for all z outside $B(M) = \{z : |z| \leq M\}$ and $\bigcup_{j=1}^{n} S_j^* \subset B(M)$. So $B(M) - \bigcup_{j=1}^{n} S_j^*$ is a nonempty closed and bounded set. Let

$$\gamma = \min_{z \in B(M) - \bigcup_{j=1}^{n} S_j^*} |f(z)|.$$

It is clear that $\gamma > 0$. Let

$$S' = \{t \in \mathbf{C}' : |t| < \min\{\gamma, 1\}\}.$$

Then S' is an open set in V and $w \in S'$. Let

$$S_j = (f|S_j^*)^{-1}(S') \quad \text{for} \quad j = 1, \cdots, n.$$

We have:

(1) $f^{-1}(S') = \bigcup_{j=1}^{n} S_j$, and all S_j are open sets in U and don't intersect each other;

(2) $f|S_j : S_j \to S'$ is a diffeomorphism.

Hence, (U, f) is a covering space of V.

Notice that $D_R \subset V$. Consider the identity $i : D_R \to V$ defined by $i(x) = x$ for all $x \in D_R$. f and i induce respectively the homomorphisms

$i_* : \pi_1(D_R, 0) \to \pi_1(V, 0)$ and $f_* : \pi_1(U, 0) \to \pi_1(V, 0)$. But the fundamental group of the disk D_R is trivial. Hence, as a subset of $\pi_1(V, 0)$,

$$i_*(\pi_1(D_R, 0)) = \{e\} \subset f_*(\pi_1(U, 0)).$$

And so by Mapping-lifting Theorem, there is a unique mapping

$$g : (D_R, 0) \to (U, 0)$$

satisfying $f \circ g = i$.

Finally, since f is analytic, g is also analytic, and the result thus follows. ¶

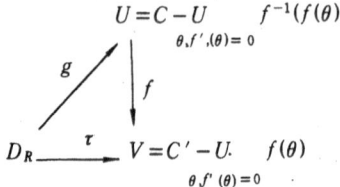

§3. One Step of Newton Iteration

Now, we will give an estimate on the effect of one iteration of the modified Newton method. The main result is Theorem 3.2.

The next theorem is a variant of Theorem 2.5.

Theorem 3.1 *Let f be a polynomial and $z \in \mathbf{C}$ with $f'(z) \neq 0$. Then there is a critical point θ of f such that for $k = 2, 3, \cdots$*

$$\left| \frac{f^{(k)}(z)}{k! f'(z)^k} \right|^{1/(k-1)} \leq \frac{K}{|f(z) - f(\theta)|}.$$

If in addition $f(z) \neq 0$ and

$$h_0 = \min_{\theta, f'(\theta) = 0} \frac{1}{K} \left| \frac{f(\theta) - f(z)}{f(z)} \right|,$$

then for $k = 2, 3, \cdots$

$$\left| \frac{f^{(k)}(z) f(z)^{k-1}}{k! f'(z)^k} \right|^{1/(k-1)} \leq \frac{1}{h_0}.$$

Proof. Expand f around z so that

$$f(v) = \sum_{k=0}^{n} \frac{f^{(k)}(z)}{k!}(v - z)^k.$$

Define the polynomial g by

$$g(u) = \sum_{k=1}^{n} \frac{f^{(k)}(z)}{k!}u^k.$$

So $g(v - z) = f(v) - f(z)$ and $g^{(k)}(v - z) = f^{(k)}(v)$, $g^{(k)}(0) = f^{(k)}(z)$ for $k = 1, 2, \cdots$. Since $f'(z) \neq 0$, applying Theorem 2.5 to g, we find that there exists a critical point σ of g such that

$$\left| \frac{f^{(k)}(z)}{k!\, f'(z)} \right|^{1/(k-1)} \left| \frac{g(\sigma)}{f'(z)} \right| \leq K, \quad k = 2, 3, \cdots.$$

Let $\theta = z + \sigma$. Then $g(\sigma) = f(\theta) - f(z)$ and $f'(\theta) = g'(\sigma) = 0$. Hence, θ is a critical point of f, and we have

$$\left| \frac{f^{(k)}(z)}{k!\, f'(z)^k} \right|^{1/(k-1)} \cdot |f(\theta) - f(z)| \leq K, \quad k = 2, 3, \cdots.$$

This gives the first part of the theorem.

The second part is a direct consequence of the first part. ¶

Theorem 3.2 *Let f be a polynomial and $z \in \mathbf{C}$ satisfying $f(z) \neq 0$ and $f'(z) \neq 0$. Let*

$$h_0 = \min_{\theta, f'(\theta)=0} \frac{1}{K} \left| \frac{f(\theta) - f(z)}{f(z)} \right|$$

and fix an $h \in (0, h_0]$. Let $z' = z - hf(z)/f'(z)$. Then there exists some $\alpha \in \mathbf{C}$ with $|\alpha| \leq 1$ such that

$$\frac{f(z')}{f(z)} = 1 - h - \frac{\alpha h^2}{h_0 - h}.$$

Proof. As in the proof of Theorem 3.1, expanding f around z and setting $z' = z - hf(z)/f'(z)$, we have

$$f(z') = \sum_{k=0}^{n} \frac{f^{(k)}(z)}{k!}(z' - z)^k$$

and

$$\frac{f(z')}{f(z)} = \sum_{k=0}^{n} \frac{f^{(k)}(z)}{k! f(z)} \left(-\frac{hf(z)}{f'(z)} \right)^k$$

$$= 1 - h + \sum_{k=2}^{n} (-h)^k \frac{f^{(k)}(z) f(z)^{k-1}}{k! f'(z)^k}$$

$$= 1 - h - \gamma h,$$

where

$$\gamma = \sum_{k=2}^{n} (-h)^{k-1} \frac{f^{(k)}(z) f(z)^{k-1}}{k! f'(z)^k}.$$

Then Theorem 3.1 shows that

$$\left| \frac{f^{(k)}(z) f(z)^{k-1}}{k! f'(z)^k} \right| \le \frac{1}{h_0^{k-1}}.$$

So

$$|\gamma| \le \sum_{k=2}^{n} |h/h_0|^{k-1} < \sum_{k=1}^{n} |h/h_0|^k$$

$$= \sum_{k=0}^{n} |h/h_0|^k - 1 = 1/(1 - h/h_0) - 1$$

$$= h_0/(h_0 - h) - 1 = h/(h_0 - h).$$

Taking $\alpha = \dfrac{(h_0 - h)\gamma}{h}$, we have

$$\frac{f(z')}{f(z)} = 1 - h - \frac{\alpha h^2}{h_0 - h},$$

and $|\alpha| \le |\gamma| |(h_0 - h)/h| \le 1$. ¶

Recall the definition of approximate zeros, the result of Theorem 3.2 is similar to the formula in the definition. The following several results are the corollaries of Theorem 3.2. Due to their importance, they are still written as theorems.

Theorem 3.3 *Let $c \ge 1$ and $\rho_f = \rho = \min_{\theta, f'(\theta)=0} |f(\theta)|$ for any polynomial f. If $f(z) \ne 0$ and $|f(z)| < \rho/(cK + K + 1)$ for some $z \in \mathbf{C}$, then*

$$|f(z')/f(z)| < 1/c,$$

where $z' = z - f(z)/f'(z)$.

Proof. Since $|f(z)| < \rho/(cK + K + 1)$ guarantees that $f'(z) \neq 0$, $z' = z - f(z)/f'(z)$ is well defined.

Notice that for every critical point θ of f,

$$|f(\theta) - f(z)| \geq |f(\theta)| - |f(z)| \geq \rho - |f(z)|.$$

By the assumptions of the theorem, $|f(z)|(cK + K + 1) < \rho$. Hence,

$$|f(z)|K(1 + c) < \rho - |f(z)| \leq |f(\theta) - f(z)|.$$

Thus

$$\frac{|f(\theta) - f(z)|}{K|f(z)|} > 1 + c.$$

By the definition of h_0 (cf. Theorem 3.2), $h_0 > 1 + c$, so $1/(h_0 - 1) < 1/c$. Then Theorem 3.2 with $h = 1$ shows that $f(z')/f'(z) = -\alpha/(h_0 - 1)$. Therefore

$$|f(z')/f(z)| < 1/c.¶$$

This theorem gives an estimate of the effect of one iteration of Newton method for the initial value z satisfying $|f(z)| < \rho/(cK + K + 1)$.

Theorem 3.4   *Suppose $|f(z_0)| < \rho/(2K + 1)$. Then the iterations of Newton method starting at z_0 converge to some z^* satisfying $f(z^*) = 0$.*

Proof. Since $|f(z_0)| < \rho/(2K + 1)$, we can choose $c > 1$ such that $|f(z_0)| < \rho/(cK + K + 1)$.

Define inductively $z_k = z_{k-1} - f(z_{k-1})/f'(z_{k-1}), k = 1, 2, \cdots$. By Theorem 3.3, for $k = 1, 2, \cdots$, we have inductively

$$|f(z_k)| < |f(z_{k-1})|/c < |f(z_0)| < \rho/(2K + 1).$$

So by the definition of ρ, $f'(z_k) \neq 0$. Hence all z_k are well defined and $|f(z_k)/f(z_0)| < (1/c)^k, k = 1, 2, \cdots$.

If for some l with $0 \leq l < \infty$ we have $f(z_0) \neq 0, f(z_1) \neq 0, \cdots$ and $f(z_{l-1}) \neq 0$, but $f(z_l) = 0$, then $z_k = z_l$ for all $k \geq l$. Therefore, $\lim_{k \to \infty} z_k = z^* = z_l$ and $f(z^*) = 0$.

If for all $k = 0, 1, 2, \cdots, f(z_k) \neq 0$ then $|f(z_k)| < |f(z_0)|/c^k$ for $k = 1, 2, \cdots$ guarantee $\lim_{k \to \infty} f(z_k) = 0$.

Now let $L = \min_{|f(z)| \leq \rho/(cK+K+1)} |f'(z)|$. Since

$$\{z : |f(z)| \leq \rho/(cK + K + 1)\}$$

is compact and $|f(z)| < \rho/(cK + K + 1) < \rho$, we have $L > 0$. Then for any positive integers k and m,

$$
\begin{aligned}
|z_{k+m} - z_k| &\leq \sum_{l=1}^{m} |z_{k+l} - z_{k+l-1}| \\
&= \sum_{l=1}^{m} |f(z_{k+l-1})/f'(z_{k+l-1})| \\
&\leq \frac{1}{L} \sum_{l=1}^{m} |f(z_{k+l-1})| \\
&\leq \frac{1}{L} \sum_{l=1}^{m} |f(z_0)|/c^{k+l-1} \\
&< \frac{|f(z_0)|}{L} \cdot c^k \left(1 - \frac{1}{c}\right).
\end{aligned}
$$

So $\{z_k\}$ is a Cauchy sequence in $\mathbf{C}$. Hence $\{z_k\}$ converges to some point $z^* \in \mathbf{C}$. Now, the continuity of f implies

$$
f(z^*) = \lim_{k \to \infty} f(z^k) = 0.
$$

This proves the theorem. ¶

Combining Theorems 3.3 and 3.4, we obtain

Theorem 3.5   *If* $|f(z)| < \rho_f/(3K+1)$ *then* z *is an approximate zero of* f.

Proof.   Applying Theorems 3.4 and 3.3 with $c = 2$, the result is obvious. ¶

It is noteworthy that Theorems 3.3, 3.4 and 3.5 give a new criterion for the feasibility of Newton method.

§4.   Conditions for Approximate Zeros

In this section we concentrate on discussing the conditions of arriving at an approximate zero after a certain number of Newton iterations. The main result is Theorem 4.6 which concerns the modified Newton method.

Lemma 4.1   (a) *If* τ, $\beta \in \mathbf{C}$ *and* $\beta \neq 0$, $\tau + \beta \neq 0$ *then*

$$
\left| \sin \arg \frac{\beta + \tau}{\beta} \right| \leq \left| \frac{\tau}{\beta} \right|;
$$

(b) If $f(z) \neq 0$ and $f(\theta) \neq 0$ then

$$\left| \sin \arg \frac{f(\theta)}{f(z)} \right| \leq \frac{|f(z) - f(\theta)|}{|f(z)|};$$

(c) If $h_* \leq 1$, $0 < h < h_*/2$, and $\alpha \in \mathbf{C}$, $|\alpha| \leq 1$, then

$$\left| \sin \arg \left(1 - h + \frac{\alpha h^2}{h_* - h} \right) \right| \leq \left(\frac{h^2}{h_* - h} \right) \left(\frac{1}{1 - h} \right).$$

Proof. (a) If τ parallels β as a vector, then $\arg((\beta + \tau)/\beta) = 0$. The result is obvious. If τ doesn't parallel β, then $|\beta|, |\tau|$ and $|\tau + \beta|$ are the three sides of some triangle, where $|\tau|$ is the opposite side of the angle formed by $|\tau|$ and $|\tau + \beta|$. Hence

$$\left| \sin \arg \frac{\beta + \tau}{\beta} \right| \leq \left| \frac{\tau}{\beta} \right|;$$

(b) Let $\beta = f(z) \neq 0$ and $\tau = f(\theta) - f(z)$. The result follows directly from (a).

(c) Since $0 < h < h_*/2$,

$$1 - h > h = \frac{h^2}{2h - h} > \frac{h^2}{h_* - h} \geq \frac{|\alpha| h^2}{h_* - h}.$$

Hence

$$\left| \sin \arg \left(1 - h + \frac{\alpha h^2}{h_* - h} \right) \right| \leq \left| \frac{\alpha h^2}{h_* - h} \right| \Big/ (1 - h)$$

$$\leq \left(\frac{h^2}{h_* - h} \right) \left(\frac{1}{1 - h} \right). \quad \P$$

Definition 4.2   Let f be a polynomial and $z_0 \in \mathbf{C}$ satisfying $f(z_0) \neq 0$. Also let $\rho_f = \min_{\theta, f'(\theta)=0} |f(\theta)| > 0$. Define

$$\mathcal{K} = \mathcal{K}_{f, z_0} = \min_{\substack{\theta, f'(\theta)=0 \\ |f(\theta)| \leq 2|f(z_0)|}} \left\{ 1, \left| \arg \frac{f(\theta)}{f(z_0)} \right| \right\},$$

$$\zeta = \zeta_{f, z_0} = \max\{2, (3K + 1)|f(z_0)|/\rho_f\}.$$

Lemma 4.3 *Let f be a polynomial and $z_0 \in \mathbf{C}$ satisfying $f(z_0) \neq 0, \rho_f > 0$ and $\mathcal{K} > 0$. Let also $h_* = \dfrac{1}{\mathcal{K}} \sin \dfrac{\mathcal{K}}{2}$. Suppose that the integer s and the real number $h > 0$, both depending on $\mathcal{K}$ and ζ, satisfy*

$$\left(1 - h + \frac{h^2}{h_* - h} \right)^s < \frac{1}{\zeta}, \tag{I}$$

and

$$s \left(\frac{h^2}{h_* - h} \right) \left(\frac{1}{1 - h} \right) < \sin \frac{\mathcal{K}}{2}. \tag{II}$$

Then for all $k = 1, 2, \cdots$, $z_k = z_{k-1} - h f(z_{k-1}) / f'(z_{k-1})$ are well defined and $|f(z_s)| < \rho_f / (3K + 1)$.

Proof. Notice that if $\rho_f > 0$ and $\mathcal{K} > 0$ then $f'(z_0) \neq 0$, and $0 < \mathcal{K} \leq 1$ implies $0 < \sin(\mathcal{K}/2) < \sin \mathcal{K} < 1$. Let θ be a critical point of f.

When $|f(\theta)| > 2|f(z_0)|$ or $|\arg(f(\theta)/f(z_0))| > \pi/2$, it is obvious that

$$\sin \left(\frac{\mathcal{K}}{2} \right) < 1 < \left| \frac{f(\theta) - f(z_0)}{f(z_0)} \right|.$$

When $|f(\theta)| \leq 2|f(z_0)|$ and $|\arg(f(\theta)/f(z_0))| \leq \pi/2$. Lemma 4.1(b) gives

$$\sin \left(\frac{\mathcal{K}}{2} \right) < \sin \mathcal{K} \leq \sin \left| \arg \frac{f(\theta)}{f(z_0)} \right| \leq \left| \frac{f(\theta) - f(z_0)}{f(z_0)} \right|.$$

Hence, with arbitrary θ, we have

$$h_* = \frac{1}{\mathcal{K}} \sin \frac{\mathcal{K}}{2} \leq h_0 = \frac{1}{\mathcal{K}} \min_{\theta, f'(\theta)=0} \left| \frac{f(z_0) - f(\theta)}{f(z_0)} \right|.$$

Now suppose $k \leq s$, and suppose inductively that for $i < k$, $f'(z_i) \neq 0$. Then $z_{i+1} = z_i - h f(z_i) / f'(z_i)$ is well defined, and moreover,

$$h_i = \frac{1}{\mathcal{K}} \min_{\theta, f'(\theta)=0} \left| \frac{f(z_i) - f(\theta)}{f(z_i)} \right| > h_*.$$

To finish the inductive proof, we prove that $f'(z_k) \neq 0$ and $h_k > h_*$.

It is easy to see that if $f(z_k) = 0$ or $f(z_i) = 0$ for some $i < k$ then the lemma is true. So we must only discuss the case $f(z_k) \neq 0$.

Notice that

$$\frac{f(z_k)}{f(z_0)} = \frac{f(z_k)}{f(z_{k-1})} \cdot \cdots \cdot \frac{f(z_1)}{f(z_0)}.$$

From Theorem 3.2 we have

$$\frac{f(z_i)}{f(z_{i-1})} = 1 - h + \frac{\alpha_i h^2}{h_{i-1} - h}, \ |\alpha_i| \le 1, \ 0 < h < h_{i-1}, \ i = 1, 2, \cdots, k.$$

By the inductive hypothesis, $0 < h < h_*/2 < h_{i-1}$, $i = 1, \cdots, k$, and thus

$$\left| 1 - h + \frac{\alpha_i h^2}{h_{i-1} - h} \right| \le |1 - h| + \left| \frac{h^2}{h_{i-1} - h} \right|$$

$$= 1 - h + \frac{h^2}{h_{i-1} - h}$$

$$< 1 - h + \frac{h^2}{h_* - h}$$

$$< 1, \quad i = 1, \cdots, k.$$

Therefore

$$\left| \frac{f(z_k)}{f(z_0)} \right| \le \left(1 - h + \frac{h^2}{h_* - h} \right)^k < 1. \tag{III}$$

On the other hand, we have

$$\arg \frac{f(z_k)}{f(z_0)} = \sum_{i=1}^{k} \arg \frac{f(z_i)}{f(z_{i-1})} \pmod{2\pi}.$$

Since

$$\left| \sin \arg \frac{f(z_i)}{f(z_{i-1})} \right| = \left| \sin \arg \left(1 - h + \frac{\alpha_i h^2}{h_{i-1} - h} \right) \right|$$

$$\le \frac{1}{1 - h} \cdot \frac{h^2}{h_{i-1} - h}$$

$$< \frac{1}{1 - h} \cdot \frac{h^2}{h_* - h}, \quad i = 1, \cdots, k,$$

we have

$$\left| \sin \arg \frac{f(z_k)}{f(z_0)} \right| \le \sum_{i=1}^{k} \left| \sin \arg \frac{f(z_i)}{f(z_{i-1})} \right|$$

$$< \; k\left(\frac{1}{1-h}\right)\left(\frac{h^2}{h_* - h}\right)$$

$$\leq \; s\left(\frac{1}{1-h}\right)\left(\frac{h^2}{h_* - h}\right)$$

$$< \; \sin\frac{\mathcal{K}}{2}.$$

Then by Lemma 2.1.1 we obtain

$$\left|\arg\frac{f(z_k)}{f(z_0)}\right| \leq \sum_{i=1}^{k}\left|\arg\frac{f(z_i)}{f(z_{i-1})}\right|$$

$$= \sum_{i=1}^{k}\left|\arg\left(1 - h + \frac{\alpha_i h^2}{h_{i-1} - h}\right)\right|$$

$$\leq \frac{\pi}{2}\sum_{i=1}^{k}\left|\frac{\alpha_i h^2}{(1-h)(h_{i-1} - h)}\right|$$

$$\leq \frac{\pi}{2}k\frac{h^2}{(1-h)(h_* - h)}$$

$$\leq \frac{\pi}{2}s\frac{h^2}{(1-h)(h_* - h)}$$

$$< \frac{\pi}{2}\sin\frac{\mathcal{K}}{2} < \frac{\pi}{2}.$$

Therefore

$$\left|\arg\frac{f(z_k)}{f(z_0)}\right| < \frac{\mathcal{K}}{2}. \qquad\qquad (IV)$$

Now, let θ be a critical point of f. We discuss the only two possible cases.

(a)   When $|f(\theta)| \leq 2|f(z_0)|$, by the definition of $\mathcal{K}$, $\left|\arg\frac{f(\theta)}{f(z_0)}\right| \geq \mathcal{K}$. If $\left|\arg\frac{f(\theta)}{f(z_k)}\right| \leq \frac{\mathcal{K}}{2}$ then $\arg\frac{f(\theta)}{f(z_0)} = \arg\frac{f(\theta)}{f(z_k)} + \arg\frac{f(z_k)}{f(z_0)}$, and thus $\left|\arg\frac{f(\theta)}{f(z_0)}\right| < \mathcal{K}$. This contradicts $\left|\arg\frac{f(\theta)}{f(z_0)}\right| \geq \mathcal{K}$. Hence

$$\left|\arg\frac{f(\theta)}{f(z_k)}\right| > \frac{\mathcal{K}}{2}. \qquad\qquad (V)$$

Notice that $|f(z_k)| < |f(z_0)| < 2|f(z_0)|$. If $f'(z_k) = 0$ then

$$\left| \arg \frac{f(z_k)}{f(z_0)} \right| \geq \mathcal{K}.$$

This contradicts (IV). So $f'(z_k) \neq 0$.

Now, if $\left| \arg \dfrac{f(\theta)}{f(z_k)} \right| \leq \dfrac{\pi}{2}$ then (V) implies

$$\left| \sin \arg \frac{f(\theta)}{f(z_k)} \right| > \sin \frac{\mathcal{K}}{2}.$$

Lemma 4.1(b) gives

$$\left| \frac{f(z_k) - f(\theta)}{f(z_k)} \right| \geq \left| \sin \arg \frac{f(\theta)}{f(z_k)} \right| > \sin \frac{\mathcal{K}}{2}.$$

If $\left| \arg \dfrac{f(\theta)}{f(z_k)} \right| > \dfrac{\pi}{2}$ then it is obvious that

$$\left| \frac{f(z_k) - f(\theta)}{f(z_k)} \right| > 1 > \sin \frac{\mathcal{K}}{2},$$

(b)   When $|f(\theta)| > 2|f(z_0)|$, $|f(z_k)| < |f(z_0)|$ means

$$\left| \frac{f(z_k) - f(\theta)}{f(z_k)} \right| > \frac{|f(\theta)| - |f(z_0)|}{|f(z_0)|} > 1 > \sin \frac{\mathcal{K}}{2}.$$

Combining (a) and (b), we have that for any critical point θ,

$$\left| \frac{f(z_k) - f(\theta)}{f(z_k)} \right| > \sin \frac{\mathcal{K}}{2}.$$

Therefore,

$$h_k = \min_{\theta, f'(\theta) = 0} \frac{1}{K} \left| \frac{f(z_k) - f(\theta)}{f(z_k)} \right| > \frac{1}{K} \sin \frac{\mathcal{K}}{2} = h_*.$$

This finishes the induction.

Combining (V) and (III), we have

$$\left| \frac{f(z_s)}{f(z_0)} \right| \leq \left(1 - h + \frac{h^2}{h_* - h} \right)^s < \frac{1}{\zeta},$$

and thus by the definition of ζ,

$$|f(z_s)| < \frac{|f(z_0)|}{\zeta} \le \frac{\rho_f}{3K+1}.$$

Finally, for $k > s$, the above discussion implies $|f(z_k)| < |f(z_s)|$, and then $f'(z_k) \ne 0$. This proves that the modified Newton method is well defined and $|f(z_k)| < \rho_f/(3K+1)$ for all $k = 1, 2, \cdots$. ¶

Now, we try to solve (I) and (II) for h and s. We need the following two lemmas.

Lemma 4.4 *If $0 < y < 1$ then $\ln(1/(1-y)) > y$.* ¶

Lemma 4.5 *If $0 < h_* < 1/2$, $\zeta > 1$ and $K \ge 1$, $c \ge 3 + \dfrac{\ln\zeta}{Kh_*}$, then*

$$\frac{c}{h_*}\left(\frac{c-1}{c-2}\right)\ln\zeta + 1 < K(c-h_*)(c-1).$$

Proof. Since $c > 3$, we have

$$\frac{\ln\zeta}{Kh_*} + \frac{c-2}{k}\frac{1}{c}\left(\frac{1}{c-1} + Kh_*\right) \quad < \quad \frac{\ln\zeta}{Kh_*} + \frac{1}{2K} + h_*$$

$$< \quad \frac{\ln\zeta}{Kh_*} + 1 \le c - 2,$$

and thus

$$\begin{aligned}
\frac{c}{h_*}\left(\frac{c-1}{c-2}\right)\ln\zeta + 1 &= \left(\frac{\ln\zeta}{Kh_*} + \frac{1}{Kc}\frac{c-2}{c-1}\right)Kc\frac{c-1}{c-2} \\
&< \left(c - 2 - \frac{h_*}{c}(c-2)\right)Kc\frac{c-1}{c-2} \\
&= K(c-h_*)(c-1). \text{¶}
\end{aligned}$$

As usual, let $\mathbf{R}^+$ be the set of positive real numbers and $\mathbf{Z}^+$ be the set of positive integers.

Theorem 4.6 *There exist two functions $H : \mathbf{R}^+ \times \mathbf{R}^+ \to \mathbf{R}^+$ and $S : \mathbf{R}^+ \times \mathbf{R}^+ \to \mathbf{Z}^+$, presented below, with the following properties: If f and z_0 satisfy $\rho_f > 0$, $K = K_{f,z_0} > 0$, $h = H(K,\zeta)$ and $s = S(K,\zeta)$ then $z_k = z_{k-1} - hf(z_{k-1})/f'(z_{k-1})$ is well inductively defined for all k and $|f(z_s)| < \rho_f/(3K+1)$. Moreover, H and S have the forms*

$$H(K,\zeta) = \frac{1}{K}\frac{\sin^2(K/2)}{3\sin(K/2) + \ln\zeta}$$

and

$$S(\mathcal{K}, \zeta) = \left\lceil K \left(3 + \frac{\ln \zeta}{\sin(\mathcal{K}/2)}\right)^2 \right\rceil.$$

Proof. By Lemma 4.3, it suffices to solve (I) and (II) for h and s. Recall that

$$\left(1 - h + \frac{h^2}{h_* - h}\right)^s < \frac{1}{\zeta}, \qquad \text{(I)}$$

$$s \frac{1}{1 - h} \frac{h^2}{h_* - h} < \sin\left(\frac{\mathcal{K}}{2}\right). \qquad \text{(II)}$$

It is obvious that (I) is equivalent to

$$\frac{\ln \zeta}{\ln(1 - h + h^2/(h_* - h))^{-1}} < s \qquad \text{(I')}$$

Let

$$y = h - \frac{h^2}{h_* - h} = h \cdot \frac{h_* - 2h}{h_* - h}.$$

Then $0 < y < 1$. But $0 < y < 1$ implies $\ln(1 - y)^{-1} > y$, thus

$$\frac{\ln \zeta}{y} < s \qquad \text{(I'')}$$

means $\dfrac{\ln \zeta}{\ln(1 - y)^{-1}} < \dfrac{\ln \zeta}{y} < s$, i.e., (I').

Let $h = h_*/c$, where $c > 2$. Then

$$y = \frac{(h_*/c)(h_* - 2h_*/c)}{h_* - h_*/c} = \left(\frac{h_*}{c}\right)\left(\frac{c - 2}{c - 1}\right),$$

and thus (I'') is equivalent to

$$\ln \zeta \cdot \left(\frac{c}{h_*}\right) \frac{c - 1}{c - 2} < s. \qquad \text{(I''')}$$

On the other hand, since $h_* = \dfrac{1}{K} \sin(\mathcal{K}/2)$, $\sin(\mathcal{K}/2) = Kh_*$, (II) is equivalent to

$$s \; < Kh_*(1 - h) \frac{h_* - h}{h^2}$$

$$= Kh_* \left(1 - \frac{h_*}{c}\right)\left(h_* - \frac{h_*}{c}\right) \bigg/ \left(\frac{h_*}{c}\right)^2$$

$$= K(c - 1)(c - h_*). \qquad \text{(II')}$$

Therefore, if we take $h = h_*/c$ then (I''') and (II') can be substituted for (I) and (II). But Lemma 4.5 says that as long as $c \geq 3 + \dfrac{\ln \zeta}{Kh_*}$ there does exist a positive integer s satisfying (I''') and (II').

Now take $c = 3 + (\ln \zeta)/Kh_*$ and $h = h_*/c$. As for s, since it does exist, we can take s big enough, for example, let $s = \lceil Kc^2 \rceil$. Now,

$$
\begin{aligned}
h &= \frac{h_*}{3 + \ln \zeta / Kh_*} = \frac{Kh_*^2}{3Kh_* + \ln \zeta} \\
&= \frac{K(1/K)^2 \sin^2(\mathcal{K}/2)}{3 \sin(\mathcal{K}/2) + \ln \zeta} \\
&= \frac{\sin^2(\mathcal{K}/2)}{K(3 \sin(\mathcal{K}/2) + \ln \zeta)},
\end{aligned}
$$

$$
s = \left\lceil K \left(3 + \frac{\ln \zeta}{\sin(\mathcal{K}/2)} \right)^2 \right\rceil.
$$

This gives the result. ¶

The theorem shows that for given f and a point z_0 satisfying $\rho_f > 0$ and $\mathcal{K} > 0$, let

$$
h = \frac{\sin^2(\mathcal{K}/2)}{K(3 \sin(\mathcal{K}/2) + \ln \zeta)}.
$$

Then for all $k = 1, 2, \cdots$, $z_k = z_{k-1} - hf(z_{k-1})/f'(z_{k-1})$ are well defined and $|f(z_s)| < \rho_f/(3K + 1)$. That is, starting at z_0, the modified Newton method with step size h can locate an approximate zero of f within s steps.

Chapter 4

A Complexity Comparison of Kuhn's Algorithm

and Newton Method

This chapter will provide a comparison between the cost estimate of Kuhn's algorithm and that of Newton method in computing approximate 1 of polynomials. The existing estimate of the former is much better than the one of the latter, and the ratio is about $n^3 \ln(n/\mu)$ to n^9/μ^7, where n is the degree of the polynomial concerned and $\mu \in (0,1)$ is the probability allowing the corresponding estimates to fail.

However, the estimate for Kuhn's algorithm is a complexity result of worst-case analysis while the one for Newton method is a result of average-case analysis. In the field of computational complexity theory for numerical methods, it is a common understanding that the handing of average-case analysis is much more difficult.

§1.   Smale's Work on the Complexity of Newton Method

Based on Chapter 3, the main idea of the Smale-founded theory of computational complexity of Newton method can be described as follows (cf. [Smale, 1981]).

Whether a point can serve as an approximate zero is dependent on the polynomial given. So in Theorem 3.4.6, the conditions on both f and z_0 are required: for an appropriate h, starting at z_0, the modified Newton method will certainly reach an approximate zero of f in s steps. In this sense, for a given polynomial, we say that a point is "good" if the relationship of the point and the polynomial satisfies the conditions of Theorem 3.4.6. Conversely, for a given point, we say that a polynomial is "good" if the relationship of the polynomial and the point satisfies the conditions. This consideration is the starting point of Smale's theory.

Let us fix $z_0 = 0$ in the rest of this chapter. Let $\mathcal{P}_n$ be the space of monic polynomials of degree n. From $\mathcal{P}_n$ we extract a set W_* which consists of "bad" polynomials with respect to the initial value $z_0 = 0$, that is, if $f \in W_*$, then no matter $h > 0$ is how small, there is no guarantee that starting at $z_0 = 0$, the modified Newton method with h as the step size will reach an approximate zero of f.

If we remove an "open" neighborhood Y_σ of W_* from $\mathcal{P}_n$, then, for all remaining polynomials , one can obtain hopefully some uniform estimates of h and s such that for every such a polynomial, the modified Newton method with h as the step size starting at $z_0 = 0$ will locate an approximate zero of the polynomial within s steps. Of course, the bigger Y_σ is, the better the remaining polynomials. The size of Y_σ will be measured by certain volume of $\mathcal{P}_n$. If the ratio of the volume of Y_σ to the volume of $\mathcal{P}_n$ is μ, $0 < \mu < 1$, and we obtain a uniform conclusion $\mathcal{A}$ for all polynomials in $\mathcal{P}_n - Y_\sigma$, then one can say that the conclusion $\mathcal{A}$ is true for any polynomial in $\mathcal{P}_n$ with probability $1 - \mu$.

According to this idea, S. Smale reported the following result: Given an integer $n > 0$ and a real number $\mu \in (0,1)$, for monic polynomials of degree n with coefficients whose absolutes are less than or equal to 1, the following is true with probability at least $1 - \mu$: starting at $z_0 = 0$, the modified Newton method with h as the step size will find an approximate zero of the polynomial in $s = \lceil (100(n + 2))^2/\mu^7 \rceil$ steps .

Concretely, we explain this as follows.

Let $\mathcal{P}_n$ be the set of monic complex polynomials of degree n. For a monic polynomial

$$f(z) = z^n + a_1 z^{n-1} + \cdots + a_n,$$

we identify $f(z)$ with the vector $a = (a_1, \cdots, a_n)$. In this sense, $\mathcal{P}_n$ is equivalent to the complex space $\mathbf{C}^n = \{(a_1, \cdots, a_n) : a_i \in \mathbf{C}\}$.

Let $P(R) = \{a \in \mathcal{P}_n : |a_i| < R, i = 1, \cdots, n\}$. It is easy to see that $\mathrm{vol}P(R)$, the volume of the polycylinder $P(R)$, is $(\pi R^2)^n$. Here we use the standard volume on $\mathbf{C}^n = \mathbf{R}^{2n}$ for $\mathcal{P}_n$ (the usual Lebesgue measurement).

Let

$$W_0 = \{f \in \mathcal{P}_n : f(\theta) = 0, f'(\theta) = 0 \text{ for some } \theta \in \mathbf{C}\}.$$

Obviously, W_0 is the set of polynomials with multiple 1. Finally, define

$$U_\rho(W_0) = \bigcup_{f_0 \in W_0} U_\rho(f_0),$$

the neighborhood of W_0, where

$$U_\rho(f_0) = \{f \in \mathcal{P}_n : |f(0) - f_0(0)| < \rho, f'(z) = f_0'(z) \text{ for all } z \in \mathbf{C}\}.$$

Notice that the above definition of the neighborhoods is different from the usual definition of neighborhoods in $\mathbf{C}^n$. In fact, $f \in U_\rho(f_0)$ if and only if $|f_0(0) - f(0)| < \rho$ and the corresponding coefficients of f and f_0 coincide except the constant terms.

Theorem 1.1 $\text{vol}[U_\rho(W_0) \cap P(R)]/\text{vol}P(R) \le (n-1)\rho^2/R^2.$ ¶

This theorem gives an estimate for the volume of $U_\rho(W_0)$ in the poly-cylinder $P(R)$.

Let

$$W_* = \{f \in \mathcal{P}_n : \text{Im}(\overline{f(0)}f(\theta)) = 0 \text{ for some critical point } \theta \text{ of } f\},$$

where $\text{Im}z$ and $\text{Re}z$ are respectively the imaginary and real part of a complex number z. It is clear that $f \in W_*$ if and only if $\text{Re}f'$, $\text{Im}f'$ and $\text{Im}(\overline{f(0)}f)$ have a common zero. Let $U_\rho(W_*)$ be a neighborhood of W_* specially defined by

$$U_\rho(W_*) = \bigcup_{f_0 \in W_*} U_\rho(f_0),$$

where as above

$$U_\rho(f_0) = \{f \in \mathcal{P}_n : |f(0) - f_0(0)| < \rho, f'(z) = f_0'(z) \text{ for all } z \in \mathbf{C}\}.$$

Then we have

Theorem 1.2 $\text{vol}(U_\rho(W_*) \cap P(R))/\text{vol}P(R) \le 3\rho(n-1)^2/R.$ ¶

Let

$$W_1 = \{f \in \mathcal{P}_n : f(\theta) = f(0) \text{ for some critical point } \theta \text{ of } f\}.$$

Define

$$L_\rho(W_1) = \bigcup_{f_0 \in W_1} L_\rho(f_0),$$

where

$$L_\rho(f_0) = \{f \in \mathcal{P}_n : |f'(0) - f_0'(0)| < \rho, |f''(0)/2 - f_0''(0)/2| < \rho\}.$$

Theorem 1.3 $\mathrm{vol}(L_\rho(W_1) \cap P(R))/\mathrm{vol}P(R) \leq 4(n+2)(\rho/R)^2$. ¶

The above two theorems give the volume estimates of $U_\rho(W_*)$ and $L_\rho(W_1)$ in the polycylinder $P(R)$.

Let

$$Q_\sigma = \{f \in \mathcal{P}_n : |f(\theta) - f(0)| < \sigma \text{ for some critical point } \theta \text{ of } f\}$$

and

$$Y_\sigma = Q_\sigma \bigcup U_{\sigma R}(W_*).$$

From Theorems 1.2 and 1.3, one can prove

Theorem 1.4   (1) *If $R > 1/3$ then*

$$\mathrm{vol}(Y_\sigma \cap P(R))/\mathrm{vol}P(R) \leq 150(n+2)^{4/3}\sigma^{2/3}.$$

(2) *If $0 < \sigma < 1$ and $f \notin Y_\sigma$ then*

(a) $\rho_f > \sigma R$;

(b) $4R \sin \mathcal{K}_f \geq \sigma^2$, *where $\mathcal{K}_f$ is $\mathcal{K}_{f,0}$ as shown in the previous chapter since $z_0 = 0$.* ¶

Finally, combining Theorems 1.4 and 3.4.6, S. Smale proved the following theorem.

Theorem 1.5   *Suppose that $\mu \in (0,1)$ and n are given and $R > 1/3$. Let $\sigma = (\mu/150)^{3/2}/(n+2)^2$. Then for every $f \in P(R)$, the following is true with probability at least $1 - \mu$: For a suitable choice of h depending on μ and n, and $z_0 = 0$, the iteration $z_k = z_{k-1} - hf(z_{k-1})/f'(z_{k-1})$ is well defined for all k, and z_s is an approximate zero of f, where*

$$s = \left\lceil 4 \left(3 + \frac{8R\log(15/\sigma)}{\sigma^2} \right)^2 \right\rceil. \P$$

Notice that for every $f \notin Y_\sigma$, by Theorems 1.4(2) and 3.4.6, the fact shown in Theorem 1.5 is true, that is, the modified Newton method will find an approximate zero of f in s steps. But the choice of σ and Theorem 1.4(1) imply $\mathrm{vol}(Y_\sigma \cap P(R))/\mathrm{vol}P(R) \leq \mu$. So, for all $f \in P(R)$, the fact shown in Theorem 1.5 is true with probability at least $1 - \mu$. It is noteworthy that, when n and μ are given, the theorem is uniformly true for all polynomials of degree n.

In particular, when $R = 1$, after some calculations, it leads to

Theorem 1.6　*There exists a function $h = h(n, \mu)$ such that for n and μ given with $0 < \mu < 1$, the following is true with probability at least $1 - \mu$ for all monic polynomials with the absolutes of all their coefficients at most 1 : Let $z_0 = 0$. Then the iteration $z_k = z_{k-1} - hf(z_{k-1})/f'(z_{k-1})$ is well defined for all $k = 1, 2, \cdots$, and z_s is an approximate zero of f, where $s = \lceil 100(n + 2))^9 / \mu^7 \rceil$.* ¶

Newton method is a widely-used and widely-studied algorithm. Now, Smale's work points out a promising way to discuss the efficiency or the computational complexity of the algorithm. In the following sections, using a way similar to Smale's, we discuss the conditions and the cost for Kuhn's algorithm to find approximate zeros for polynomials. With the results for Kuhn's algorithm in Chapter 2, the discussion here is successful. The cost estimate for Kuhn's algorithm in finding approximate zeros is better than the one for Newton method.

Section 2 is extracted from [Smale, 1981]. Section 3 discusses relevant problems on applying Kuhn's algorithm to find approximate zeros.

§2.　Set of Bad Polynomials and its Volume Estimate

Let $\mathcal{P}_n$, $P(R)$, W_0 and $U_\rho(W_0)$ be as in Section 1. This section gives some properties of $U_\rho(W_0)$.

As in Section 3.3, we define $\rho_f = \min_{\theta, f'(\theta)=0} |f(\theta)|$. Thus $f \in W_0$ if and only if $\rho_f = 0$. This motivates to give

Theorem 2.1　$f \in U_\rho(W_0)$ *if and only if* $\rho_f < \rho$.

Proof.　It suffices to prove that $f \in U_\rho(W_0)$ if and only if $f(\theta) < \rho$ for some critical point θ.

Suppose $f \in U_\rho(W_0)$. By definition of $U_\rho(W_0)$, there exists some $f_0 \in W_0$ such that $f \in U_\rho(f_0)$. By definition of $U_\rho(f_0)$, the absolute of the difference between the constant terms of f and f_0 is at most ρ, and the other coefficients of f coincide with those of f_0. Thus, for all $z \in \mathbf{C}$, $|f(z) - f_0(z)| < \rho$. Since $f_0 \in W_0$, there is a critical point θ of f_0 such that $f_0(\theta) = 0$. Notice that $f_0'(z) = f'(z)$ for all $z \in \mathbf{C}$. Hence, θ is also a critical point of f and thus $|f(\theta)| = |f(\theta) - f_0(\theta)| < \rho$.

Now suppose that $|f(\theta)| < \rho$ for some critical point θ of f. Define a polynomial f_0 by $f_0(z) = f(z) - f(\theta)$ for all $z \in \mathbf{C}$. It is obvious that θ is a critical point of f_0. But $|f_0(\theta)| = |f(\theta) - f(\theta)| = 0$. So $f_0 \in W_0$ and

$f \in U_\rho(f_0)$. Therefore, $f \in U_\rho(W_0)$. ¶

The above theorem provides us with a clear structure of $U_\rho(W_0)$. The rest of this section will give an estimate for the volume of $U_\rho(W_0)$.

Next theorem can be found in [Jacobson, 1974].

Theorem 2.2   Let $f(z) = \sum_{j=0}^{n} a_{n-j} z^j$ and $g(z) = \sum_{j=0}^{m} b_{m-j} z^j$ with $n > 0$ and $m > 0$. Denote by $R(f, g)$ the resultant of f and g. Then $R(f, g) = 0$ if and only if $a_0 = b_0 = 0$ or $f(z)$ and $g(z)$ have a nontrivial common factor. ¶

For example, let $f(z) = az^2 + bz + c$ with $a \neq 0$ and take $g(z) = 2az + b$. Then

$$R(f, f') = \begin{vmatrix} a & b & c \\ 2a & b & 0 \\ 0 & 2a & b \end{vmatrix} = a(4ac - b^2).$$

Thus, we obtain a well-known fact: f has a multiple zero if and only if $4ac - b^2 = 0$.

Lemma 2.3   The subset $W_0 \subset \mathcal{P}_n$ has a representation as a complex algebraic hypersurface given by a polynomial equation $F(a) = 0$, where $F : \mathbf{C}^n \to \mathbf{C}$ is defined by

$$F(a_1, \cdots, a_n) = \sum_{j=0}^{n-1} F_j(a_1, \cdots, a_{n-1}) a_n^j,$$

all F_j are polynomials and F has a total degree of $2n - 1$.

In other words, the theorem says

$$W_0 = \{a \in \mathbf{C}^n : F(a) = 0\}.$$

Proof.   Notice that $f \in W_0 \iff f$ and f' have a common zero $\iff$ $f(z)$ and $f'(z)$ have a common factor $\iff R(f, f') = 0$.

Using the notations of Theorem 2.2, $g = f'$, $m = m - 1$, $a_0 = 1$ and $b_0 = n$. The resultant $R(f, f')$ is a polynomial in $a_1, \cdots, a_n$ with total degree $2n - 1$, and its degree in a_n is $n - 1$. Let $F = R(f, f')$. The lemma follows. ¶

Theorem 2.4 $\mathrm{vol}[U_\rho(W_0) \cap P(R)] / \mathrm{vol} P(R) \leq (n-1)\rho^2/R^2$.

This theorem has been introduced as Theorem 1.1 in Section 1.

Proof.   Define $\chi : \mathbf{C}^n \to \mathbf{R}$ to be the characteristic function of $U_\rho(W_0)$, that is, χ is 1 in $U_\rho(W_0)$ and zero otherwise.

Notice that $W_0 = \{a \in \mathbf{C}^n : F(a) = 0\}$, where the degree of F in a_n is $n - 1$. Thus for almost every $(a_1, \cdots, a_{n-1})$, the intersection of W_0 with the one (complex) dimensional coordinate line through $(a_1, \cdots, a_{n-1})$ consists of at most $n - 1$ points. For every intersection point, there is an open neighborhood consisting of the points whose distances to the one (complex) dimensional line along a_n is at most ρ. Then

$$\left| \int_{|a_n|<\rho} \chi(a_1, \cdots, a_n) da_n \right| \leq \left| \int_{|a_n|<+\infty} \chi(a_1, \cdots, a_n) da_n \right|$$
$$\leq (n-1)\pi\rho^2.$$

Hence, by Fubini Theorem (cf. [Royden, 1968]), we have

$$\frac{\text{vol}[U_\rho(W_0) \cap P(R)]}{\text{vol}P(R)}$$
$$= \frac{1}{(\pi R^2)^n} \int_{\rho(R)} \chi(a) da$$
$$= \frac{1}{(\pi R^2)^n} \int_{|a_1|<R} \cdots \int_{|a_{n-1}|<R} \left[\int_{|a_n|<R} \chi(a_1, \cdots, a_n) da_n \right] da_{n-1} \cdots da_1$$
$$\leq \frac{1}{(\pi R^2)^n} \int_{|a_1|<R} \cdots \int_{|a_{n-1}|<R} (n-1)\pi\rho^2 da_{n-1} \cdots da_1$$
$$= \frac{1}{(\pi R^2)^n} (n-1)\pi\rho^2 (\pi R^2)^{n-1}$$
$$= \frac{(n-1)\rho^2}{R^2}.$$

This completes the proof. ¶

§3. Locate Approximate Zeros by Kuhn's Algorithm

We have shown in Chapters 1 and 2 that for every complex polynomial, using Kuhn's algorithm, one can find all zeros of a polynomial within a given accuracy in a finite number of complementary pivots, and we also give an estimate of the number of the complementary pivots needed. It is known that, Kuhn's algorithm itself does not requires the concept of approximate zeros. But for making an "exact" comparison between the cost estimates for the two algorithms, we intentionally consider applying Kuhn's algorithm to find approximate zeros.

Recall that Theorem 2.4 in Chapter 2 asserts that with Kuhn's algorithm one can find all zeros of a polynomial within a given accuracy, and the theorem gives also the corresponding cost estimate. The nearer a point approaches a zero of a polynomial, the smaller the absolute value of the polynomial at the point. Theorem 3.5 in Chapter 3 affirms that if $|f(z)| < \rho_f/(3K+1)$ for some z then z is an approximate zero of f. But it is noteworthy that $|f(z)| < \rho_f/(3K+1)$ is of no meanings for the polynomial f with multiple zeros. Moreover, to obtain a uniform estimate of computational speeds, we must remove a neighborhood of the set of polynomials with multiple zeros from the polynomial space to evade the cases of $\rho_f = 0$ and of too small ρ_f. This is the main idea of this section.

On the other hand, Theorem 2.1 shows that $f \in U_\rho(W_0)$ if and only if $\rho_f < \rho$, that is, $U_\rho(W_0) = \{f \in \mathcal{P}_n : \rho_f < \rho\}$. Theorem 2.4 shows that $\mathrm{vol}(U_\rho(W_0) \cap P(R))/\mathrm{vol}P(R) \le (n-1)\rho^2/R^2$. Thus we are in a position to present the following theorem.

Theorem 3.1   *Given a positive integer n and a positive real number σ, for any $f \in P(R)$ and $f \notin U_{\sigma R}(W_0)$, Kuhn's algorithm can locate n approximate zeros for all the n zeros of f in at most s steps, where*

$$s = \lceil 5\pi M^2 + 28\pi n(1+0.75n)^2 \lceil \log_2(\sqrt{2}(1+0.75n)/\epsilon)\rceil \rceil$$

with

$$M = \max\left\{\frac{3\sqrt{2}(2+\pi)n}{4\pi}, 1 + \frac{5nR}{4(n-1)}\right\} + \sqrt{2}$$

and

$$\epsilon = \frac{\sigma(1-M)^2}{13[1-(n+1)M^n + nM^{n+1}]}.$$

Proof.   When $\sigma R \le \rho_f$. Let $z \in \mathbf{C}$ satisfy $|z-\xi| < \epsilon$ for some zero ξ of f, then

$$
\begin{aligned}
|f(z)| &= |f(z) - f(\xi)| \\
&= \left|\sum_{j=0}^{n} a_{n-j}(z^j - \xi^j)\right| \\
&\le |z-\xi|\sum_{j=1}^{n} |a_{n-j}(z^{j-1} + z^{j-2}\xi + \cdots + \xi^{j-1})| \\
&\le \epsilon R \sum_{j=1}^{n} j M^{j-1}
\end{aligned}
$$

$$\begin{aligned}
&= \ \epsilon R \left(\sum_{j=1}^{n} x^j \right)'_M \\
&= \ \epsilon R \left(\frac{x - x^{n+1}}{1 - x} \right)'_M \\
&= \ \epsilon R \frac{1 - (n+1)M^n + nM^{n+1}}{(1 - M)^2} \\
&\leq \ \frac{\sigma R}{13} \\
&\leq \ \frac{\rho_f}{13} \\
&\leq \ \frac{\rho_f}{(3K + 1)}.
\end{aligned}$$

Therefore z is an approximate zero of f.

When $\sigma R \geq \rho_f$, Theorem 2.4 in Chapter 2 shows that Kuhn's algorithm can find n approximate zeros for all the n zeros of f in s steps. But $U_{\sigma R}(W_0) = \{f \in \mathcal{P}_n : \rho_f < \sigma R\}$, hence, $f \notin U_{\sigma R}(W_0)$ if and only if $\rho_f \geq \sigma R$. This finishes the proof. ¶

Theorem 3.2 *Suppose that $\mu \in (0,1)$ and n are given. Then for any $f \in P(R)$, Kuhn's algorithm can find, with probability at least $1 - \mu$, n approximate zeros for the all n zeros of f in at most s steps, where*

$$s = \lceil 5\pi M^2 + 28\pi n(1 + 0.75n)^2 \lceil \log_2(\sqrt{2}(1 + 0.75n)/\epsilon) \rceil \rceil$$

with

$$M = \max \left\{ \frac{3\sqrt{2}(2 + \pi)n}{4\pi}, 1 + \frac{5nR}{4(n-1)} \right\} + \sqrt{2}$$

and

$$\epsilon = \frac{\mu^{1/2}(1 - M)^2}{13n^{1/2}[1 - (n+1)M^n + nM^{n+1}]}.$$

Proof. Let $\sigma = (\mu/n)^{1/2}$. Theorem 3.1 shows that for any $f \in P(R)$ with $f \notin U_{\sigma R}(W_0)$, Kuhn's algorithm can locate n approximate zeros for all the n zeros of f within s steps. Here

$$\begin{aligned}
\epsilon \ &= \ \frac{\sigma(1 - M)^2}{13[1 - (n+1)M^n + nM^{n+1}]} \\
&= \ \frac{\mu^{1/2}(1 - M)^2}{13n^{1/2}[1 - (n+1)M^n + nM^{n+1}]}.
\end{aligned}$$

Moreover, by Theorem 2.4

$$\mathrm{vol}(U_{\sigma R}(W_0) \bigcap P(R))/\mathrm{vol}P(R) = (n-1)\sigma^2 R^2/R^2 = (n-1)\mu/n < \mu.$$

Therefore, for every $f \in P(R)$, by Kuhn's algorithm one can find n approximate zeros within s steps with probability at least $1 - \mu$. ¶

Recall Smale's result for Newton method: Given a positive integer n and $\mu \in (0,1)$, for every $f \in P(R)$, with a suitable choice of h, by the modified Newton method one can reach an approximate zero of f in $s = \lceil (100(n+2))^9/\mu^7 \rceil$ steps with probability at least $1 - \mu$. Comparing it with Theorem 3.2, by means of some careful calculations, we know that as for finding approximate 1 for Newton method, the estimate of Kuhn's algorithm is much better than Smale's estimate on Newton method. In fact, notice that in the sense of probability, the cost of finding one approximate zero is , in average, $1/n$ of the cost of locating n approximate zeros. Thus we obtain that

Corollary 3.3 Given an integer $n > 0$ and $\mu \in (0,1)$, for every $f \in P(R)$, averagely speaking, using Kuhn's algorithm one can find an approximate zero of f in $s = \lceil 140(n+2)^3 \log_2(n/\mu) \rceil$ steps with probability at least $1 - \mu$. ¶

This gives the ratio $n^3 \ln(n/\mu)$ to n^9/μ^7 shown at the beginning of this chapter.

§4. Some Remarks

As for cost estimates, $n^3 \ln(n/\mu)$ is, of course, much better than n^9/μ^7 if only the presentations of the final results are concerned. However, one should by no means underestimate the great significance of Smale's pioneering work on Newton method. In fact, we deal with Kuhn's algorithm with a nice geometrical and topological structure while S. Smale discusses Newton method which is much more difficult to handle.

Traditionally, convergence is usually the first consideration for any algorithm. If there has been certain guarantee of its convergence, then one can begin to discuss its efficiency, or its computational complexity. As it was seen in the previous chapters, Kuhn's algorithm is the one with a strong guarantee of global convergence. On the other hand, Newton method is even not convergent in the worst case, and is very slowly convergent in some worse cases. It seams that if one does not know whether the computation is convergent or not, how can he or she discuss the cost of the computation?

In his work, S. Smale excludes resolutely and sophisticatedly the worst case and some worse cases, and so leads to a wonderful result of probabilistic estimate. Though the appearance of $n^3 \ln(n/\mu)$ is better than the one of n^9/μ^7, we should remember that they have very different origins. Perhaps Fig. 4.1 provides some ideas to evaluate the significances of works of n^9/μ^7 and of $n^3 \ln(n/\mu)$. It is a very informal figure. Let f denote polynomials and c the costs of computing approximate zeros of the polynomials. The thin lines mean the real cost, and the thick lines the cost estimates. Thus the left figure represents Smale's work on Newton method with the small wiggles as his exclusion of the worst and the worse cases, and the right one is the work on Kuhn's algorithm.

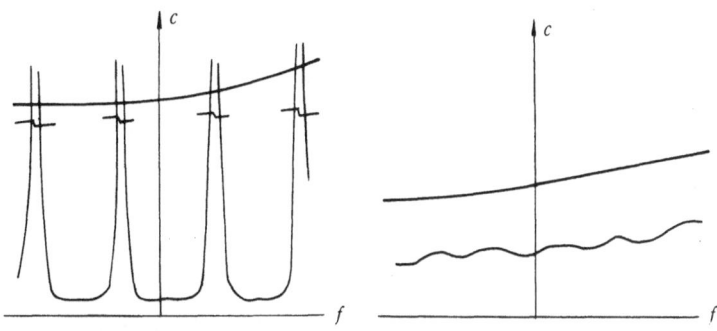

Figure 4.1

It is a great challenge to develop the computational complexity when there is even no general guarantee of computational convergence. The key point is the introduction of the probability term μ. The development from the worst-case analysis to the average-case analysis, or the development from the absolute cost estimate to the probabilistic cost estimate in the computational complexity theory, is, in some sense, logically a further continuation of the development from the generality consideration to the genericity consideration in mathematics. As it is well-known in applied mathematics, there are some numerical methods usually working very well but casually very poor. If insisting on discussing from the worst-case view point, they should be all considered as "bad" or "poor" methods. It would lead to a confrontation between theory and practice. Thus the development of average-case analysis shows us a future of reasonable and harmonized development in the computational complexity theory.

Just along this path, [Smale, 1983] proves G. Dantzig's conjecture that when using the simplex method to solve a randomly chosen linear programming problem with a fixed number of constraints, the number of needed iterations grows linearly with the number of variables. Some scholars believe that Dantzig's simplex method for linear programming is a method of scientific computation with greatest economical benefits in this century.

On the other hand, the pioneering work on the computational complexity of Newton method does not have nothing to be desired. From the view point of constructive mathematics, the way designed by Smale using the modified Newton iterations to locate nice initial values called approximate zeros is still not entirely constructive in the thoroughgoing sense. One may think that this is a blemish in an otherwise totally perfect thing. The point is at the selection of the step size h. Yes, Smale provides a formula of h depending on both ζ and $\mathcal{K}$. Both ζ and $\mathcal{K}$ themselves are in turn depending on ρ_f, the least one of the absolutes of the critical values of f. Notice that the critical points of $f(z)$ are the zeros of $f'(z)$, the derivative of the original polynomial. Thus in order to locate a nice initial value to start Newton iteration for the zeros of the polynomial $f(z)$, one should first find out the zeros of another polynomial $f'(z)$. It shows some logical difficulty to be entirely of thoroughgoing constructiveness.

We guess that it will be a problem of more difficulty. Perhaps a combination of Kuhn's algorithm and Newton method is a way to make the discussion entirely constructive: Kuhn's algorithm is used for approximate zeros, the nice initial value of Newton method. Then both the two stages are constructive.

The following chapter is still devoted to the work of S. Smale and his colleagues on the computational complexity of polynomial equations.

Chapter 5

Incremental Algorithms and Cost Theory

In Chapter 3, we got familiar with the modified Newton method $z' = z - hf(z)/f'(z)$, that is, the incremental Newton algorithm. In this chapter, based on [Shub & Smale, 1985; 1984], we will introduce the general incremental algorithm $I_{h,f}$, and, in particular, we will discuss Euler algorithm $E_{k(h,f)}$, generalized Euler algorithm $G_{k(h,f)}$ and Taylor algorithm $T_{k(h,f)}$.

After introducting several incremental algorithms in Section 1, in Section 2 we present the concept of the incremental algorithms of efficiency k, and Theorem 2.11 shows that the incremental algorithm $E_{k(h,f)}$ is of efficiency k. In Section 3, we define the concept of the generalized approximate zeros for general incremental algorithms. Theorems 4.8, 4.9, 4.10 and Theorem 5.9 give respectively the estimates of the number of iterations needed and of the cost of $E_{k(h,f)}$ for finding an approximate zero of a polynomial f. Theorem 2.15 shows that starting at an approximate zero z of f, $z_l = E^l_{k(h,f)}(z)$ tends fast to a zero of f. Theorem 6.5 is a natural extension of Theorems 4.8 and 4.9, and is true for any incremental algorithm $I_{h,f}$ of efficiency k .

§1.   Incremental Algorithms $I_{h,f}$

First, we introduce the concept of the incremental algorithms in Definition 1.1. The definition is rather general. After the definition, we will present several examples to explain the concept. The incremental algorithms given in the examples are those discussed latter in this chapter.

Definition 1.1   Let $f : C \to C$ be a polynomial and S^2 be the Riemann space of complex numbers (i.e., $S^2 = C \bigcup \{\infty\}$). By an incremental algorithm, we mean a mapping

$$I_{h,f} : S^2 \to S^2, \quad z' = I_{h,f}(z) = I(z) \quad \text{with} \quad I_{0,f}(z) = z$$

parameterized by $0 < h \leq 1$ and the complex polynomial f.

To solve $f(z) = 0$ with an incremental algorithm $I_{h,f}$ we have to choose appropriately an initial value z_0 and h such that the sequence

$$z_m = I_{h,f}(z_{m-1}) = I_{h,f}^m(z_0), \quad m = 1, 2, \cdots$$

converges to a solution of $f(z) = 0$. The concept of incremental algorithms is derived from some standard iterative processes for solving either non-linear systems of equations or ordinary differential equations. Now, we give several examples.

Example 1.2 (Incremental Newton Algorithm)(cf. Chapter 3)

$$I_{h,f} = N_{h,f}(z) = z - h\frac{f(z)}{f'(z)}, \quad f'(z) \neq 0.$$

The special case $N(z) = N_{1,f}(z) = z - f(z)/f'(z)$ when $h = 1$ is the original Newton method.

Example 1.3 (Incremental Euler Algorithm) Given a polynomial f and a complex number z, we define $r(f, z)$ as follows. If $f'(z) \neq 0$ then

$$r = r(f, z) = \min_{\theta, f'(\theta)=0} |f(z) - f(\theta)| = |f(z) - f(\theta_*)|,$$

where $f'(\theta_*) = 0$. If $f'(z) = 0$ then $r(f, z) = 0$.

When $f(z) \neq 0$, define

$$
\begin{aligned}
h_1 &= h_1(f, z) \\
&= \frac{r(f, z)}{|f(z)|} \\
&= \min_{\theta, f'(\theta)=0} \frac{|f(z) - f(\theta)|}{|f(z)|} \\
&= \frac{|f(z) - f(\theta_*)|}{|f(z)|},
\end{aligned}
$$

where $\theta_* = \theta_*(f, z)$ is a critical point of f, that is, $f'(\theta_*) = 0$.

Now, suppose that $r = r(f, z) > 0$. By Lemma 3.2.3, the convergent radius of the inverse f_z^{-1} of f is at least $r(f, z)$. Moreover, the analytic function

$$f_z^{-1} : D_r(f(z)) \to \mathbf{C}$$

sends $f(z)$ to z, where $D_r(f(z)) = \{w \in \mathbf{C} : |w - f(z)| < r\}$. When $h < h_1(f, z)$,

$$|hf(z)| < h_1(f, z)|f(z)| = \min_{\theta, f'(\theta)=0} |f(z) - f(\theta)| = r.$$

Thus the equation

$$\frac{f(z')}{f(z)} = 1 - h$$

is solvable and its solution is $z' = f_z^{-1}((1-h)f(z))$. Expanding the inverse around $f(z)$, we obtain the incremental Euler algorithm:

$$
\begin{aligned}
z' &= E_\infty(z) \\
&= E_{\infty(h,f)}(z) \\
&= f_z^{-1}(f(z')) \\
&= f_z^{-1}((1 - h)f(z)) = f_z^{-1}(f(z) - hf(z)) \\
&= f_z^{-1}(f(z)) + \sum_{l=1}^{\infty} \frac{(f_z^{-1})_{f(z)}^{(l)}}{l!}(-hf(z))^l \\
&= z + \sum_{l=1}^{\infty} \frac{(f_z^{-1})_{f(z)}^{(l)}}{l!}(-hf(z))^l.
\end{aligned}
$$

It is obvious that $E_{\infty(0,f)}(z) = z$.

The above iteration is usually uncomputable in practice, so we need truncate the latest power series. For this, let T_k be the operation of truncating a power series starting at the $(k+1)$-th term, that is,

$$T_k\left(\sum_{l=0}^{\infty} a_l h^l\right) = \sum_{l=0}^{k} a_l h^l.$$

Then we obtain the following k-th incremental Euler algorithm:

$$
\begin{aligned}
E_k(z) &= E_{k(h,f)}(z) \\
&= T_k f_z^{-1}((1 - h)f(z)) \\
&= z + \sum_{l=1}^{k} \frac{(f_z^{-1})_{f(z)}^{(l)}}{l!}(-hf(z))^l \\
&= z + \frac{f(z)}{f'(z)}\left[\sum_{l=1}^{k} \frac{(f_z^{-1})_{f(z)}^{(l)}(-1)^{l-1}}{l!} f^{l-1}(z)f'(z)h^l\right],
\end{aligned}
$$

Obviously, $E_{k(0,f)}(z) = z$ for $k = 1, 2, \cdots$.

It is easy to see that $E_1(z) = z - hf(z)/f'(z)$, a special incremental Euler algorithm, is exactly the incremental Newton algorithm.

For the incremental Euler algorithms shown above, we have

Theorem 1.4 *Let*

$$\sigma_i = (-1)^{i-1} \frac{f^{(i)}(z) f^{i-1}(z)}{i! (f'(z))^i}.$$

Then

$$E_1(z) = z - \frac{f(z)}{f'(z)};$$

$$E_2(z) = z - \frac{f(z)}{f'(z)}(h - \sigma_2 h^2);$$

$$E_3(z) = z - \frac{f(z)}{f'(z)}[h - \sigma_2 h^2 + (2\sigma_2^2 - \sigma_3)h^3];$$

$$E_4(z) = z - \frac{f(z)}{f'(z)}[h - \sigma_2 h^2 + (2\sigma_2^2 - \sigma_3)h^3 - (5\sigma_2^3 - 5\sigma_2\sigma_3 + \sigma_4)h^4];$$

Proof. Let $w = f(z)$ and $z = f^{-1}(w)$. Then

$$(f^{-1})'(w) = \frac{1}{f'(z)},$$

$$(f^{-1})'' = -\frac{f'' \cdot (f^{-1})'}{(f')^2} = -\frac{f''}{(f')^3},$$

$$(f^{-1})''' = \frac{-f'''(f')^3 + 3(f')^2 \cdot (f'')^2}{(f')^6} \cdot \frac{1}{f'} = -\frac{f'''}{(f')^4} + \frac{3(f'')^2}{(f')^5}.$$

On the other hand,

$$\sigma_2 \frac{f}{f'} = (-1)^{2-1} \frac{f'' \cdot f^{2-1}}{2(f')^2} \cdot \frac{f}{f'} = -\frac{f'' \cdot (-f)^2}{2!(f')^3} = \frac{1}{2!}(f^{-1})''(-f)^2,$$

$$-(2\sigma_2^2 - \sigma_3)\frac{f}{f'} = -\frac{f}{f'}\left\{ 2\left[\frac{f'' \cdot f}{2!(f')^2}\right]^2 - \left[\frac{f''' \cdot f^2}{3!(f')^3}\right]^2 \right\}$$

$$= \frac{1}{3!}\frac{-f'''(f')^3 + 3(f')^2(f'')^2}{(f')^7}(-f)^3$$

$$= \frac{1}{3!}(f^{-1})'''(-f)^3,$$

$$(5\sigma_2^3 - 5\sigma_2\sigma_3 + \sigma_4)\frac{f}{f'}$$

$$= \frac{f}{f'}\left[-5\left(\frac{f''\cdot f}{2!(f')^2}\right)^3 + 5\left(\frac{f''\cdot f}{2(f')^2}\right)\left(\frac{f'''\cdot f^2}{3!(f')^3}\right) - \frac{f^{(4)}\cdot f^3}{4!(f')^4}\right]$$

$$= -\frac{f^{(4)}\cdot f^4}{4!(f')^5} + 4\cdot\frac{f'''\cdot f^4\cdot f''}{4!(f')^6} + 6\cdot\frac{f''\cdot f'''\cdot f^4}{4!(f')^6} - 15\cdot\frac{(f'')^3\cdot f^4}{4!(f')^7}$$

$$= \frac{f^4}{4!}\cdot\frac{1}{f'}\left[\frac{-f^{(4)}\cdot(f')^4 + 4f'''\cdot(f')^3\cdot f''}{(f')^8}\right.$$

$$\left. + \frac{6f''\cdot f'''\cdot(f')^5 - 15(f'')^2\cdot(f')^4\cdot f''}{(f')^{10}}\right]$$

$$= \frac{1}{4!}(f^{-1})^{(4)}(-f)^4.$$

From the above equations, we have

$$E_4(z) = z + \sum_{l=1}^{4}\frac{(f_z^{-1})_{f(z)}^{(l)}}{l!}(-hf(z))^l$$

$$= z - \frac{f(z)}{f'(z)}[h - \sigma_2 h^2 + (2\sigma_2^2 - \sigma_3)h^3 - (5\sigma_2^3 - 5\sigma_2\sigma_3 + \sigma_4)h^4].$$

Example 1.5 (Generalized Incremental Euler Algorithm) Suppose that $c_1, \cdots, c_k$ are real numbers given with $c_1 > 0$. Let

$$P(h) = c_1 h + c_2 h^2 + \cdots + c_k h^k.$$

Set h to be small enough so that $|P(h)| < h_1$. Then

$$|P(h)f(z)| < h_1|f(z)| = \min_{\theta, f'(\theta)=0}|f(z) - f(\theta)| = r,$$

where h_1 and r are as shown in Example 1.3. With an argument similar to Example 1.3, solving the equation

$$\frac{f(z')}{f(z)} = 1 - P(h)$$

gives the generalized incremental Euler algorithm

$$z' = G_\infty(z)$$
$$= G_{\infty(h,f)}(z)$$
$$= f_z^{-1}((1 - P(h))f(z))$$

$$= f_z^{-1}(f(z) - P(h)f(z))$$

$$= z + \sum_{l=1}^{\infty} \frac{(f_z^{-1})_{f(z)}^{(l)}}{l!}(-P(h)f(z))^l.$$

Obviously, $G_{\infty(0,f)}(z) = z$.

Similarly, by truncating the above power series, we obtain the k-th generalized incremental Euler algorithm as follows.

$$G_k(z) = G_{k(h,f)}(z)$$

$$= T_k f_z^{-1}((1 - P(h))f(z))$$

$$= z + T_k \sum_{l=1}^{\infty} \frac{(f_z^{-1})_{f(z)}^{(l)}}{l!}(-P(h)f(z))^l$$

$$= z + \sum_{l=1}^{k} \frac{(f_z^{-1})_{f(z)}^{(l)}}{l!}(-1)^l f^l(z) \cdot T_k \left(\sum_{j=1}^{k} c_j h^j\right)^l$$

$$= z + \sum_{j=0}^{k-1} \left[\sum_{l=1}^{k} \frac{(f_z^{-1})_{f(z)}^{(l)}}{l!}(-1)^l f^l(z) \sum_{\substack{t_1 + \cdots + t_l = j+1 \\ 1 \le t_s \le k}} c_{t_1} \cdots c_{t_l}\right] h^{j+1}$$

$$= z - \frac{f(z)}{f'(z)} \sum_{j=0}^{k-1}$$

$$\left[\sum_{l=1}^{k} \sum_{\substack{t_1 + \cdots + t_l = j+1 \\ 1 \le t_s \le k}} c_{t_1} \cdots c_{t_l}(-1)^{l-1} \frac{(f_z^{-1})_{f(z)}^{(l)}}{l!} f^{l-1}(z)f'(z)\right] h^{j+1}$$

$$= z - \frac{f(z)}{f'(z)} \sum_{j=0}^{k-1} P_j h^{j+1},$$

where

$$P_j = \sum_{l=1}^{k} \sum_{\substack{t_1 + \cdots + t_l = j+1 \\ 1 \le t_s \le k}} c_{t_1} \cdots c_{t_l}(-1)^{l-1} \frac{(f_z^{-1})_{f(z)}^{(l)}}{l!} f^{l-1}(z)f'(z).$$

It is clear that $G_{k(0,f)}(z) = z$.

Similarly to Theorem 1.4, for the first several generalized incremental Euler algorithms, we have

Theorem 1.6 *Let*

$$P_j = \sum_{l=1}^{k} \sum_{\substack{t_1+\cdots+t_l=j+1 \\ 1 \le t_s \le k}} c_{t_1} \cdots c_{t_l} (-1)^{l-1} \frac{(f_z^{-1})_{f(z)}^{(l)}}{l!} f^{l-1}(z) f'(z).$$

Then

$$P_0 = c_1,$$
$$P_1 = c_2 - \sigma_2 c_1^2,$$
$$P_2 = c_3 - 2\sigma_2 c_1 c_2 - (\sigma_3 - 2\sigma_2^2) c_1^3.$$

Proof.

$$P_0 = c_1(-1)^0 \frac{(f_z^{-1})'}{1!} f^0(z) \cdot f'(z) = c_1,$$

$$
\begin{aligned}
P_1 &= c_2(-1)^0 \frac{(f_z^{-1})'}{1!} f^0(z) \cdot f'(z) + c_1^2 (-1)^1 \frac{(f_z^{-1})''}{2!} f(z) \cdot f'(z) \\
&= c_2 - \sigma_2 c_1^2,
\end{aligned}
$$

$$
\begin{aligned}
P_2 &= c_3(-1)^0 \frac{(f_z^{-1})'}{1!} f^0(z) \cdot f'(z) + 2c_1 c_2 (-1)^1 \frac{(f_z^{-1})''}{2!} f(z) \cdot f'(z) \\
&\quad + c_1^3 (-1)^2 \frac{(f_z^{-1})'''}{3!} f^2(z) \cdot f'(z) \\
&= c_3 - 2\sigma_2 c_1 c_2 + (2\sigma_2^2 - \sigma_3) c_1^3. \P
\end{aligned}
$$

Remark 1.7 By setting $c_1 = 1$ and $c_2 = \cdots = c_k = 0$, it is easy to see that the incremental Euler algorithm is a special case of the generalized incremental Euler algorithm.

Now, we introduce the polynomial

$$\sigma(w) = \sum_{i=1}^{n} \sigma_i w^i,$$

where

$$\sigma_i = (-1)^{i-1} \frac{f^{(i)}(z) f^{i-1}(z)}{i! (f'(z))^i}$$

and n is the degree of f. Notice that $\sigma(0) = 0$ and $\sigma'(0) = \sigma_1 = 1$. If we rewrite the incremental algorithm $z' = I_{h,f}(z)$ as

$$z' = I_{h,f}(z) = z + F(z) R(h, f, z),$$

where $F(z) = -f(z)/f'(z)$, then we have

Theorem 1.8   *Let σ, F, R be as above. Then*

$$\frac{f(z')}{f(z)} = 1 - \sigma(R), \ \text{ where } R = \frac{I_{h,f}(z) - z}{F(z)}.$$

Proof.   By Taylor formula,

$$
\begin{aligned}
f(z') &= f(z) + \sum_{i=1}^{n} \frac{f^{(i)}(z)}{i!}(z' - z)^i \\
&= f(z) + \sum_{i=1}^{n} \frac{f^{(i)}(z)}{i!} F^i R^i \\
&= f(z) + \sum_{i=1}^{n} \frac{f^{(i)}(z)}{i!} \left(-\frac{f(z)}{f'(z)}\right)^i R^i \\
&= f(z) - f(z) \sum_{i=1}^{n} \sigma_i R^i \\
&= f(z)[1 - \sigma(R)].
\end{aligned}
$$

This gives the theorem. ¶

Now, we are ready to represent the incremental Euler algorithm and the generalized incremental Euler algorithm in a simple way.

Theorem 1.9

$$G_k(z) = z + FT_k \sum_{l=1}^{\infty} \frac{(\sigma^{-1})^{(l)}(0)}{l!}(P(h))^l;$$

$$E_k(z) = z + F \sum_{l=1}^{k} \frac{(\sigma^{-1})^{(l)}(0)}{l!} h^l.$$

Proof.   Applying Theorem 1.8 to $I_{h,f} = E_{\infty(h,f)}$, we have

$$\frac{f(z')}{f(z)} = 1 - h = 1 - \sigma(R).$$

Thus $h = \sigma(R)$. Since $\sigma(0) = 0$ and $\sigma'(0) = 1$, σ is invertible to σ^{-1} on some open disk with $\sigma^{-1}(0) = 0$ and $(\sigma^{-1})'(0) = 1$. Thus we have

$$\sigma^{-1}(h) = R = \frac{I_{h,f}(z) - z}{F} = \frac{1}{F} \sum_{l=1}^{\infty} \frac{(f_z^{-1})_{f(z)}^{(l)}}{l!}(-f(z))^l h^l.$$

On the other hand, notice that $|hf(z)| < r = h_1|f(z)|$ for $|h| < h_1$. Expanding σ^{-1} around 0, we have

$$\sigma^{-1}(h) = \sum_{l=1}^{\infty} \frac{(\sigma^{-1})^{(l)}(0)}{l!} h^l.$$

It follows that

$$(\sigma^{-1})^{(l)}(0) = \frac{1}{F}(f_z^{-1})^{(l)}_{f(z)}(-f(z))^l$$

and

$$
\begin{aligned}
G_k(z) &= z + T_k \sum_{l=1}^{\infty} \frac{(f_z^{-1})^{(l)}_{f(z)}}{l!}(-f(z))^l (P(h))^l \\
&= z + FT_k \sum_{l=1}^{\infty} \frac{(\sigma^{-1})^{(l)}(0)}{l!}(P(h))^l.
\end{aligned}
$$

When $P(h) = h$, it gives

$$
\begin{aligned}
E_k(z) &= z + FT_k \sum_{l=1}^{\infty} \frac{(\sigma^{-1})^{(l)}(0)}{l!} h^l \\
&= z + F \sum_{l=1}^{k} \frac{(\sigma^{-1})^{(l)}(0)}{l!} h^l. ¶
\end{aligned}
$$

Finally, we introduce

Example 1.10 (Incremental Taylor Algorithm) Let

$$T_k(z) = T_{k(h,f)}(z) = z + \sum_{i=1}^{k} \frac{1}{i!} \frac{d^i \phi_t(z)}{dt^i}\bigg|_{t=0} h^i,$$

where $\phi_t(z)$ is the solution of the following differential equation,

$$
\begin{cases}
\dfrac{du}{dt} = -\dfrac{f(u)}{f'(u)} = F(u) \\
u|_{t=0} = z,
\end{cases}
$$

that is,

$$
\begin{cases}
\dfrac{d\phi_t(z)}{dt} = -\dfrac{f(\phi_t(z))}{f'(\phi_t(z))} = F(\phi_t(z)) \\
\phi_0(z) = z.
\end{cases}
$$

By some simple calculation, we have

$$\frac{d\phi_t(z)}{dt}\bigg|_{t=0} = -\frac{f(\phi_t(z))}{f'(\phi_t(z))}\bigg|_{t=0} = -\frac{f(z)}{f'(z)} = F(z),$$

$$\begin{aligned}
\frac{d^2\phi_t(z)}{dt^2}\bigg|_{t=0} &= F'(\phi_t(z))\frac{d\phi_t(z)}{dt}\bigg|_{t=0} \\
&= F(\phi_t(z))F'(\phi_t(z))|_{t=0} \\
&= F(z)F'(z),
\end{aligned}$$

$$\begin{aligned}
\frac{d^3\phi_t(z)}{dt^3}\bigg|_{t=0} &= \frac{d}{dt}[F(\phi_t(z))F'(\phi_t(z))]|_{t=0} \\
&= [F'^2(z) + F(z)F''(z)]F(z).
\end{aligned}$$

Thus

$$T_1(z) = z - h\frac{f(z)}{f'(z)} = z + Fh = E_1(z),$$

$$T_2(z) = z + Fh + \frac{1}{2}FF'h^2,$$

$$T_3(z) = z + Fh + \frac{1}{2}FF'h^2 + \frac{1}{6}(F'^2 + FF'')Fh^3.$$

With the above differential equations, we have

$$\frac{f'(\phi_t(z))}{f(\phi_t(z))}\frac{d\phi_t(z)}{dt} = -1,$$

$$\ln f(\phi_t(z)) = -t + \ln c(z),$$

$$f(\phi_t(z)) = c(z)e^{-t}.$$

Since $\phi_0(z) = z$,

$$f(\phi_t(z)) = f(\phi_0(z))e^{-t} = e^{-t}f(z).$$

Thus it follows that

$$\begin{aligned}
z + \sum_{i=1}^{\infty}\frac{1}{i!}\frac{d^i\phi_t(z)}{dt^i}\bigg|_{t=0} h^i \\
= \phi_h(z) = f_z^{-1}(e^{-h}f(z)) \\
= f_z^{-1}\left(f(z)\left(1 - \sum_{j=1}^{\infty}\frac{(-1)^{j-1}h^j}{j!}\right)\right) \\
= z + \sum_{l=1}^{\infty}\frac{(f_z^{-1})_{f(z)}^{(l)}}{l!}\left[-\sum_{j=1}^{\infty}\frac{(-1)^{j-1}h^j}{j!}f(z)\right]^l
\end{aligned}$$

and

$$T_k(z) \;=\; z + \sum_{i=1}^{k} \frac{1}{i!} \frac{d^i \phi_t(z)}{dt^i}\bigg|_{t=0} h^i$$

$$=\; z + T_k \sum_{l=1}^{\infty} \frac{(f_z^{-1})_{f(z)}^{(l)}}{l!} \left[-\sum_{j=1}^{\infty} \frac{(-1)^{j-1} h^j}{j!} f(z) \right]^l$$

$$=\; z + T_k \sum_{l=1}^{\infty} \frac{(f_z^{-1})_{f(z)}^{(l)}}{l!} \left[-\sum_{j=1}^{k} \frac{(-1)^{j-1} h^j}{j!} f(z) \right]^l.$$

If $c_j = (-1)^{j-1}/j!$ for $j = 1, \cdots, k$ then $T_k(z) = G_k(z)$.

Before closing this section, we give the following important concept in discussions of efficiency.

Definition 1.11 The incremental algorithm $I_{h,f}$ is of efficiency k if there exist real constants $\delta > 0$, $K > 0$, $c_1 > 0$ and $c_2, \cdots, c_k$ such that

$$\frac{f(z')}{f(z)} = \frac{f(I_{h,f}(z))}{f(z)} = 1 - (c_1 h_1 + \cdots + c_k h_k) + S_{k+1}(h),$$

and for any $0 < h \le \delta \min\{1, h_1\}$,

$$|S_{k+1}(h)| \le K h^{k+1} \max\left\{ 1, \frac{1}{h_1^k} \right\},$$

where $h_1 > 0$.

Remark 1.12 By the definition, if an incremental algorithm $I_{h,f}$ is of efficiency k, it must be of efficiency k' for all $1 \le k' \le k$. In the above examples, it is easy to know that E_∞ is of efficiency k for $k = 1, 2, \cdots$, and so is G_∞ in Example 1.5. In the next section, we will prove that E_k is also of efficiency k.

§2.   Euler's Algorithm Is of Efficiency k

First, we do some preparations.

Definition 2.1 Let $D_r = \{z : |z| < r\}$ and $f : D_1 \to \mathbf{C}$ be an analytic function given by the power series

$$f(z) = \sum_{j=1}^{\infty} a_j z^j$$

which converges on D_1. If $a_1 = 1$ and f is one-to-one on D_1, then f is called a Schlicht function.

For Schlicht functions, we have

Theorem 2.2 (Bieberbach-De Branges Theorem)   *If f is a Schlicht function, then $|a_m| \le m$ for $m = 2, 3, \cdots$.* ¶

Refer to [De Branges, 1985] for a proof. This theorem is derived from the classic Bieberbach Conjecture.

Theorem 2.3 (Bieberbach-Koebe Theorem)   *If f is a Schlicht function, then its image* $\mathrm{Image}(f) \supset D_{1/4}$. ¶

See [Hille, 1962] for proof of the theorem, .

Theorem 2.4 (Loewner Theorem)   *Suppose that $g(w) = w + b_2 w^2 + b_3 w^3 + \cdots$ is the inverse of some Schlicht function $f(z) = \sum_{j=1}^{\infty} a_j z^j$, then for $m = 1, 2, \cdots$,*

$$|b_m| \le 2^m \frac{1 \cdot 3 \cdots (2m - 1)}{1 \cdot 2 \cdots (m + 1)} \le 4^{m-1}.$$

Recall that we have encountered this theorem in Chapter 3. Otherwise see [Hayman, 1958]. Notice that $b_1 = 1$ in the representation of $g(w)$

Theorem 2.5 (Distortion Theorem of Koebe and Gronwall)   *If f is Schlicht, then for $|z| \le r$,*

$$\frac{r}{(1 + r)^2} \le |f(z)| \le \frac{r}{(1 - r)^2}$$

and

$$\frac{1 - r}{(1 + r)^3} \le |f'(z)| \le \frac{1 + r}{(1 - r)^3}.$$

Refer to [Hille, 1962] or [Hayman, 1958] for proof.

Next simple lemma is technical.

Lemma 2.6 $\sum_{l=l_0}^{\infty} l x^{l-l_0} \le l_0 / (1 - x)^2$ *for all* $0 \le x \le 1$.

Proof.

$$\sum_{l=l_0}^{\infty} l x^{l-l_0} = x^{1-l_0} \sum_{l=l_0}^{\infty} l x^{l-1}$$

$$= x^{1-l_0} \left(\sum_{l=l_0}^{\infty} x^l \right)'$$

$$= x^{1-l_0} \left(x^{l_0} \sum_{j=0}^{\infty} x^j \right)'$$

$$= x^{1-l_0} \left(\frac{x^{l_0}}{1-x} \right)'$$

$$= x^{1-l_0} \frac{l_0 x^{l_0-1}(1-x) + x^{l_0}}{(1-x)^2}$$

$$= \frac{l_0 - (l_0 - 1)x}{(1-x)^2}$$

$$\leq \frac{l_0}{(1-x)^2} \cdot \P$$

The following two Lemmas concern the properties of convergent series.

Lemma 2.7 Let $g(x) = \sum_{j=1}^{\infty} b_j x^j$ be convergent with $b_1 = 1$. Let $a \geq \max\{\max_{l \geq 2} |b_l|^{1/(l-1)}, 1\}$. Then for $|x| < 1/a$,

$$\left| \frac{g^{(l)}(x)}{l!} \right| \leq a^{l-1} \left(\frac{1}{1-a|x|} \right)^{l+1}.$$

Proof. Let $y = a|x| < 1$. Then

$$\left| \frac{g^{(l)}(x)}{l!} \right| \leq \frac{1}{l!} \left| \sum_{j=1}^{\infty} j(j-1) \cdots (j-l+1) b_j x^{j-l} \right|$$

$$\leq \frac{1}{l!} \sum_{j=1}^{\infty} j(j-1) \cdots (j-l+1) a^{j-1} |x|^{j-l}$$

$$\leq \frac{a^{l-1}}{l!} \sum_{j=1}^{\infty} j(j-1) \cdots (j-l+1) y^{j-l}$$

$$= \frac{a^{l-1}}{l!} \frac{d^l}{dy^l} \left(\sum_{j=0}^{\infty} y^j \right)$$

$$= \frac{a^{l-1}}{l!} \frac{d^l}{dy^l} \left(\frac{1}{1-y} \right)$$

$$= \frac{a^{l-1}}{(1-y)^{l+1}}$$

$$= a^{l-1} \left(\frac{1}{1-a|x|} \right)^{l+1} \cdot \P$$

Lemma 2.8   *Let $g(w) = \sum_{j=1}^{\infty} b_j w^j$ be convergent with $b_1 = 1$ and $a \geq \max\{\max_{l \geq 2} |b_l|^{1/(l-1)}, 1\}$. Let $x, w \in \mathbf{C}$ and b, c be positive numbers. Suppose that $(1+c)ab < 1, |x| \leq b$ and $|w - x| \leq bc$. Then*

$$|g(w) - g(x)| \leq \frac{bc}{(1 - ab)(1 - (1+c)ab)}.$$

Proof.   By Taylor formula and Lemma 2.7, we have

$$
\begin{aligned}
|g(w) - g(x)| &= \left| \sum_{l=1}^{\infty} \frac{g^{(l)}(x)}{l!} (w - x)^l \right| \\
&\leq \sum_{l=1}^{\infty} a^{l-1} \left(\frac{1}{1 - a|x|} \right)^{l+1} |w - x|^l \\
&\leq \sum_{l=1}^{\infty} a^{l-1} \left(\frac{1}{1 - a|x|} \right)^{l+1} (bc)^l \\
&= \left(\frac{1}{1 - a|x|} \right)^2 bc \sum_{l=1}^{\infty} \left(\frac{abc}{1 - a|x|} \right)^{l-1} \\
&= bc \left(\frac{1}{1 - a|x|} \right)^2 \frac{1}{1 - abc/(1 - a|x|)} \\
&= \frac{bc}{1 - a|x|} \cdot \frac{1}{1 - a|x| - abc} \\
&\leq \frac{bc}{(1 - ab)(1 - (1+c)ab)}. ¶
\end{aligned}
$$

Next two theorems are important for the proof of the main result of this section.

Theorem 2.9   *Let f be Schlicht and g be its inverse. Then*

$$|g(f(z)) - g(T_k f(z))| \leq \frac{r^{k+1}(1 - r)^2(k + 1)}{((1 - r)^2 - 4r)((1 - r)^2 - 4r(1 + r^k(k + 1)))},$$

where $r = |z|$ with r less than the first positive zero of the denominator

$$((1 - u)^2 - 4u)((1 - u)^2 - 4u(1 + u^k(k + 1))).$$

Proof.   Take $a = 4$, and set $b = r/(1 - r)^2$ and $c = r^k(k + 1)$. Let $f(z) = \sum_{j=1}^{\infty} a_j z^j$ with $a_1 = 1$. By Koebe-Gronwall Distortion Theorem,

$$|f(z)| \leq \frac{r}{(1 - r)^2}, \quad \text{where } r = |z|.$$

Denote $w = f(z)$ and $x = T_k f(z)$. By Lemma 2.6,

$$
\begin{aligned}
|w - x| &= |f(z) - T_k f(z)| \\
&= \left| \sum_{j=k+1}^{\infty} a_j z^j \right| \\
&\le \sum_{j=k+1}^{\infty} |a_j| r^j \\
&\le \sum_{j=k+1}^{\infty} j r^j \\
&\le \frac{(k+1) r^{k+1}}{(1-r)^2} \\
&= bc.
\end{aligned}
$$

Moreover,

$$
\begin{aligned}
(1 + c) ab &= (1 + r^k(k+1)) \frac{4r}{(1-r)^2} \\
&= 1 - \frac{(1-r)^2 - 4r(1 + r^k(k+1))}{(1-r)^2} \\
&< 1.
\end{aligned}
$$

With Lemma 2.8, we have

$$
\begin{aligned}
&\quad |g(w) - g(x)| \\
&= |g(f(z)) - g(T_k f(z))| \\
&\le \frac{bc}{(1 - ab)(1 - (1+c)ab)} \\
&= \frac{r^{k+1}(k+1)}{(1-r)^2} \Big/ \left(1 - \frac{4r}{(1-r)^2}\right)\left(1 - (1 + r^k(k+1))\frac{4r}{(1-r)^2}\right) \\
&= \frac{r^{k+1}(1-r)^2(k+1)}{((1-r)^2 - 4r)((1-r)^2 - 4r(1 + r^k(k+1)))}. \quad \blacksquare
\end{aligned}
$$

Next, we extend slightly the previous theorem to the case of $|z| < R$.

Theorem 2.10   Let $f(z) = \sum_{j=1}^{\infty} a_j z^j$ be a one-to-one analytic function on $|z| < R$ with $a_1 = 1$ and g be its inverse. Then

$$
|g(f(z)) - g(T_k f(z))| \le \frac{R r^{k+1}(1-r)^2(k+1)}{((1-r)^2 - 4r)((1-r)^2 - 4r(1 + r^k(k+1)))},
$$

where $r = |z|/R$ with r less than the first positive zero of the denominator

$$((1 - u)^2 - 4u)((1 - u)^2 - 4u(1 + u^k(k + 1))).$$

Proof. Let $w = f(z)$, $z = g(w)$, $z = R\tilde{z}$ and $w = R\tilde{w}$. Then $\tilde{w} = \tilde{f}(\tilde{z}) = f(R\tilde{z})/R$ is Schlicht, and $\tilde{z} = \tilde{g}(\tilde{w}) - g(R\tilde{w})/R$ is its inverse.

By Theorem 2.9,

$$\left| \frac{1}{R} g(f(z)) - \frac{1}{R} g(T_k f(z)) \right|$$

$$= \left| \tilde{g}\left(\frac{1}{R} f(R\tilde{z})\right) - \tilde{g}\left(T_k \frac{1}{R} f(R\tilde{z})\right) \right|$$

$$= |\tilde{g}(\tilde{f}(\tilde{z})) - \tilde{g}(T_k \tilde{f}(\tilde{z}))|$$

$$\leq \frac{r^{k+1}(1 - r)^2(k + 1)}{((1 - r)^2 - 4r)((1 - r)^2 - 4r(1 + r^k(k + 1)))},$$

that is,

$$|g(f(z)) - g(T_k f(z))| \leq \frac{Rr^{k+1}(1 - r)^2(k + 1)}{((1 - r)^2 - 4r)((1 - r)^2 - 4r(1 + r^k(k + 1)))}. \quad ¶$$

Now, we prove the main result of this section.

Theorem 2.11 *For a polynomial f and a complex number z with $f'(z) \neq 0$, $f(z) \neq 0$ and $r(f, z) > 0$, let $z' = E_k(z) = E_{k(h,f)}(z)$ for $k = 1, 2, \cdots$. Then*

$$\frac{f(z')}{f(z)} = 1 - h + Q(h, f, z)\frac{h^{k+1}}{h_1^k}, \quad |Q| \leq \beta_k(r),$$

where

$$\beta_k(r) = \frac{(k + 1)(1 - r)^2}{((1 - r)^2 - 4r)((1 - r)^2 - 4r(1 + r^k(k + 1)))}$$

with $r = h/h_1$, $h_1 = h_1(f, z)$ as defined in Example 1.3, $0 < r < r_k$ and r_k is the first positive zero of the denominator of $\beta_k(r)$.

Moreover, the k-th incremental Euler algorithm E_k is of efficiency k for $k = 1, 2, \cdots$, and $|f(z')| \leq |f(z)|$ for h small enough.

Proof. Using the polynomial $\sigma(w) = \sum_{i=1}^{n} \sigma_i w^i$ defined in Section 1 just after Remark 1.7, by Theorem 1.8, we have

$$\frac{f(z')}{f(z)} = 1 - \sigma(R),$$

where

$$R = \frac{E_{k(h,f)}(z) - z}{F}$$

$$= \sum_{l=1}^{k} \frac{(\sigma^{-1})^{(l)}(0)}{l!} h^l$$

$$= T_k \sum_{l=1}^{\infty} \frac{(\sigma^{-1})^{(l)}(0)}{l!} h^l$$

$$= T_k \sigma^{-1}(h).$$

Since σ^{-1} is a one-to-one analytic function on

$$D_{h_1} = \{u \in \mathbf{C} : |u| < h_1\},$$

by Theorem 2.10, we obtain

$$|\sigma(\sigma^{-1}h) - \sigma(T_k(\sigma^{-1}h))|$$

$$\leq \frac{h_1 r^{k+1}(1-r)^2(k+1)}{((1-r)^2 - 4r)((1-r)^2 - 4r(1+r^k(k+1)))}$$

$$= \frac{h^{k+1}}{h_1^k}\beta_k(r).$$

Hence,

$$\frac{f(z')}{f(z)} = 1 - \sigma(R) = 1 - \sigma(T_k(\sigma^{-1}h))$$

$$= 1 - h + (\sigma(\sigma^{-1}h) - \sigma(T_k(\sigma^{-1}h)))$$

$$= 1 - h + Q(h, f, z)\frac{h^{k+1}}{h_1^k}$$

and

$$|Q| \leq \beta_k(r).$$

Next, we prove that E_k is of efficiency k. For this, let $0 < \delta < r_k$ and $K = \beta_k(\delta)$, where δ is independent of h, f and z. When $0 < h \leq \delta \min\{1, h_1\} \leq \delta h_1$,

$$r = \frac{h}{h_1} \leq \delta < r_k.$$

Thus

$$|S_{k+1}(h)| = \left| Q(h, f, z)\frac{h^{k+1}}{h_1^k} \right|$$

$$\leq \frac{\beta_k(r)}{h_1^k} h^{k+1}$$

$$\leq \beta_k(\delta) h^{k+1} \max\left\{1, \frac{1}{h_1^k}\right\}$$

$$\leq K h^{k+1} \max\left\{1, \frac{1}{h_1^k}\right\},$$

where the used inequality $\beta_k(r) \leq \beta_k(\delta)$ will be proved in Lemma 2.12. ¶

Next three lemmas describe the properties of α_k, β_k and γ_k.

Lemma 2.12 *If $0 < r < r_k$ then $\beta_k'(r) > 0$. Thus $\beta_k(r)$ is strictly increasing in r.*

Proof. The direct calculation shows that

$$\begin{aligned}
\beta_k'(r) = {} & 2(k+1)(1-r)[(2+2r)((1-r)^2 - 4r(1+r^k(k+1))) \\
& + (1-r)((1-r)^2 - 4r)((1-r) + 2(1+r^k(k+1))) \\
& + 2rk(k+1)r^{k-1}]/[\cdots]^2,
\end{aligned}$$

where

$$[\cdots] = ((1-r)^2 - 4r)((1-r)^2 - 4r(1+r^k(k+1))).$$

It is obvious that, when $0 < r < r_k < 1$, all the terms in the above equation are positive. Thus $\beta_k'(r) > 0$. ¶

Lemma 2.13 *r_k increases in k and $\lim_{k\to\infty} r_k = 3 - \sqrt{8} = 0.1715\cdots$, where $3 - \sqrt{8}$ is the least positive zero of $(1-r)^2 - 4r$.*

Proof. Since $r^k(k+1) < r^{k+1}k$ for $0 < r < 3 - \sqrt{8}$, we have

$$\begin{aligned}
(1-r)^2 - 4r(1 + r^{k-1}k) &< (1-r)^2 - 4r(1 + r^k(k+1)) \\
&< (1-r)^2 - 4r.
\end{aligned}$$

Let $F_k(r) = (1-r)^2 - 4r(1 + r^k(k+1))$ for $k = 1, 2, \cdots$ and $F(r) = (1-r)^2 - 4r$. It is easy to see that

$$F_k(r) < F_{k+1}(r) < F(r), \quad F_k'(r) < 0, \text{ and } F'(r) < 0.$$

Then referring to Fig. 5.1, we see that r_k increases in k. Let $r_* = \lim_{k\to\infty} r_k$. Then

$$\begin{aligned}
0 = \lim_{k\to\infty} F_k(r_k) &= \lim_{k\to\infty} [(1-r_k)^2 - 4r_k(1 + r_k^k(k+1))] \\
&= (1-r_*)^2 - 4r_*.
\end{aligned}$$

Therefore,

$$r_* = 3 - \sqrt{8}. \, ¶$$

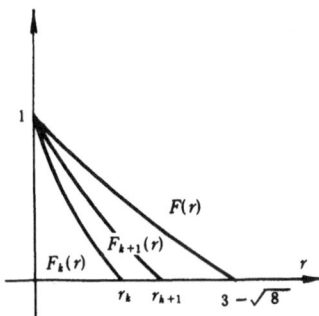

Figure 5.1

Lemma 2.14 *Let $\alpha_k(r) = \beta_k(r)r^k$ for $0 < r < r_k$. Then $\alpha'_k(r) > 0$ and $\alpha_k(r) = 1$ has a unique solution $\bar{r}_k$ which increases in k with $\lim_{k\to\infty} \bar{r}_k = 3 - \sqrt{8}$.*

Proof. Obviously, $\beta_k(r) > 0$. By Lemma 2.12, $\beta'_k(r) > 0$. Thus

$$\alpha'_k(r) = \beta'_k(r)r^k + kr^{k-1}\beta_k(r) > 0.$$

Since $\lim_{r\to 0} \alpha_k(r) = 0$, $\lim_{r\to r_k} \alpha_k(r) = +\infty$ and $\alpha_k(r)$ is strictly increasing in r, there is a unique $\bar{r}_k$ such that $\alpha_k(\bar{r}_k) = 1$. It is obvious that

$$0 < \bar{r}_k < r_k < 3 - \sqrt{8}.$$

From

$$1 = \alpha_k(\bar{r}_k) = \frac{(k+1)(1-\bar{r}_k)^2\bar{r}_k^k}{((1-\bar{r}_k)^2 - 4r_k)((1-\bar{r}_k)^2 - 4\bar{r}_k(1 + \bar{r}_k^k(k+1)))}$$

we have

$$\begin{aligned}
0 &= \lim_{k\to\infty} (k+1)(1-\bar{r}_k)^2\bar{r}_k^k \\
&= \lim_{k\to\infty} ((1-\bar{r}_k)^2 - 4r_k)((1-\bar{r}_k)^2 - 4\bar{r}_k(1 + \bar{r}_k^k(k+1))) \\
&= \lim_{k\to\infty} [(1-\bar{r}_*)^2 - 4\bar{r}_*]^2,
\end{aligned}$$

where $\bar{r}_* = \lim_{k\to\infty} \bar{r}_k = 3 - \sqrt{8}$.

Now, it suffices to show that $\lim_{k\to\infty} \bar{r}_k$ exists and is also finite, and therefore, it is sufficient to prove that $\bar{r}_k$ is strictly increasing in k. Similarly to the proof of Lemma 2.13, from $k/(k+1) \geq 1/2 > 3 - \sqrt{8} > r > 0$

we have $kr^{k-1} > (k+1)r^k$. Thus $\alpha_k(r) > \alpha_{k+1}(r)$. Since $\alpha'_k(r) > 0$, referring to Fig. 5.2, we have

$$\bar{r}_k < \bar{r}_{k+1} < 3 - \sqrt{8}.\P$$

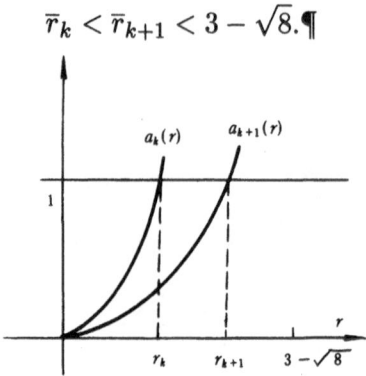

Figure 5.2

Finally, we prove the following theorem which is important for the discussion of computational complexity theory, and from which the concept of generalized approximate zeros is derived.

Theorem 2.15   *Let $\rho_f = \min_{\theta, f'(\theta)=0} |f(\theta)| > 0$. If a polynomial f and a complex number z satisfy*

$$|f(z)| = b\left(\frac{\bar{r}_k}{1 + \bar{r}_k}\right)\rho_f,$$

then with $h = 1$, $(E_k)^l(z) = z_l$ is well defined for all l and $\lim_{l \to \infty} z_l = z_*$ with $f(z_*) = 0$.

Moreover, for all $l > 0$,

$$|f(z_l)| \leq M|f(z_{l-1})|^{k+1}, \quad (convergence\ of\ order\ k+1)$$

where $M = (b/|f(z)|)^k$.

Finally, for all $l > 0$,

$$|f(z_l)| \leq b^{(k+1)^l}\frac{\bar{r}_k}{1 + \bar{r}_k}\rho_f.$$

Proof.   Let

$$r = \frac{|f(z)|}{|f(z) - f(\theta_*)|} = \frac{1}{h_1},$$

where $f'(\theta_*) = 0$. Since

$$
\begin{aligned}
|f(z) - f(\theta_*)| &\geq |f(\theta_*)| - |f(z)| \\
&\geq \rho_f - b\left(\frac{\bar{r}_k}{1 + \bar{r}_k}\right)\rho_f \\
&\geq \left(1 - \frac{\bar{r}_k}{1 + \bar{r}_k}\right)\rho_f \\
&= \frac{\rho_f}{1 + \bar{r}_k} > 0,
\end{aligned}
$$

we have

$$
\begin{aligned}
r &= \frac{h}{h_1} = \frac{1}{h_1} = \frac{|f(z)|}{|f(z) - f(\theta_*)|} \\
&< \left(\frac{b\bar{r}_k}{1 + \bar{r}_k}\right)\rho_f \Big/ \frac{\rho_f}{1 + \bar{r}_k} = b\bar{r}_k < \bar{r}_k < r_k.
\end{aligned}
$$

Thus with $h = 1$, by Theorem 2.11 and Lemma 2.12,

$$
\left|\frac{f(z')}{f(z)}\right| = \left|1 - 1 + Q\frac{1}{h_1^k}\right| = \left|Q\frac{1}{h_1^k}\right| \leq \beta_k(r)r^k < \beta_k(\bar{r}_k)\bar{r}_k = 1
$$

and

$$
\begin{aligned}
\left|\frac{f(z')}{f(z)}\right| &\leq \beta_k(r)r^k \\
&\leq \beta_k(\bar{r}_k)\bar{r}_k \\
&= \beta_k(\bar{r}_k)\frac{|f(z)|^k}{|f(z) - f(\theta_*)|^k} \\
&\leq \beta_k(\bar{r}_k)\left(\frac{1 + \bar{r}_k}{\rho_f}\right)^k |f(z)|^k \\
&= M|f(z)|^k,
\end{aligned}
$$

where

$$
M = \beta_k(\bar{r}_k)\left(\frac{1 + \bar{r}_k}{\rho_f}\right)^k = \frac{1}{\bar{r}_k^k}\left(\frac{1 + \bar{r}_k}{\rho_f}\right)^k = \left(\frac{1 + \bar{r}_k}{\bar{r}_k \rho_f}\right)^k = \left(\frac{b}{|f(z)|}\right)^k.
$$

This implies

$$
|f(z')| \leq M|f(z)|^{k+1}.
$$

Next, we prove the results by induction. Let

$$
y_0 = b\bar{r}_k, \quad y_l = (y_{l-1})^{k+1}\beta_k(\bar{r}_k), \quad z_0 = z
$$

and

$$s_0 = \frac{|f(z)|}{\min_{\theta, f'(\theta)=0} |f(z) - f(\theta)|} = \frac{1}{h_1}.$$

Consider

$$(1_l) \quad |f(z_l)| \leq y_l \frac{\rho_f}{1 + \bar{r}_k};$$

$$(2_l) \quad s_l \leq y_l,$$

$l = 0, 1, 2, \cdots,$ where

$$s_l = \frac{|f(z_l)|}{\min_{\theta, f'(\theta)=0} |f(z_l) - f(\theta)|}.$$

By the assumption of the theorem, (1_0) is true. Since $r < b\bar{r}_k$ shown above, (2_0) is also true. We will inductively prove that (1_l) implies (2_l) while (1_{l-1}) and (2_{l-1}) imply (1_l).

In fact, by (1_l) and (2_l)

$$\begin{aligned}
s_l &= \frac{|f(z_l)|}{\min_{\theta, f'(\theta)=0} |f(z_l) - f(\theta)|} \\
&\leq \left(y_l \frac{\rho_f}{1 + \bar{r}_k} \right) \frac{1}{\min_{\theta, f'(\theta)=0} |f(z_l) - f(\theta)|} \\
&\leq \left(y_l \frac{\rho_f}{1 + \bar{r}_k} \right) \frac{1 + \bar{r}_k}{\rho_f} \\
&= y_l.
\end{aligned}$$

By (1_{l-1}) and (2_{l-1}), we obtain (1_l), that is,

$$\begin{aligned}
|f(z_l)| &\leq \beta_k(\bar{r}_k) s_{l-1}^k |f(z_{l-1})| \\
&\leq \beta_k(\bar{r}_k) s_{l-1}^k y_{l-1} \frac{\rho_f}{1 + \bar{r}_k} \\
&\leq \beta_k(\bar{r}_k) y_{l-1}^k y_{l-1} \frac{\rho_f}{1 + \bar{r}_k} \\
&= y_l \frac{\rho_f}{1 + \bar{r}_k}.
\end{aligned}$$

Moreover, we can prove inductively that

$$y_0 = b^{(k+1)^0} \bar{r}_k = b\bar{r}_k,$$

$$y_{l-1} = b^{(k+1)^{l-1}} \bar{r}_k$$

and

$$
\begin{aligned}
y_l &= (y_{l-1})^{k+1}\beta_k(\overline{r}_k) \\
&= (b^{(k+1)^{l-1}}\overline{r}_k)^{k+1}\beta_k(\overline{r}_k) \\
&= b^{(k+1)^l}\beta_k(\overline{r}_k)\overline{r}_k^{k+1} \\
&= b^{(k+1)^l}\overline{r}_k \\
&< \overline{r}_k < 3 - \sqrt{8}.
\end{aligned}
$$

This proves that

$$
|f(z)| \leq y_l \frac{\rho_f}{1+\overline{r}_k} = b^{(k+1)^l}\frac{\overline{r}_k}{1+\overline{r}_k}\rho_f.
$$

For z_l, since $b_l = b^{(k+1)^l} \leq b < 1$, $(E_k)^l(z) = z_l$ is well determined for all $l > 0$.

By $|f(z')| \leq M|f(z)|^{k+1}$ proved above, we have

$$
|f(z_l)| \leq M|f(z_{l-1})|^{k+1}.
$$

Finally, $|f(z_l)| \leq b^{(k+1)^l}\dfrac{\overline{r}_k}{1+\overline{r}_k}\rho_f < \rho_f$ implies $f'(z_l) \neq 0$. Hence for any $l > 0$, $(E_k)^l(z) = z_l$ is well defined, and with a proof similar to that of Theorem 3.4 in Chapter 3, we obtain that $\lim_{k\to\infty} z_l = z_*$ with $f(z_*) = \lim_{k\to\infty} f(z_l) = 0$. ¶

§3. Generalized Approximate Zeros

Based on the previous section, we are now ready to present the concept of generalized approximate zeros.

Definition 3.1 Let f be a polynomial. We call z an approximate zero for f relative to k if $|f(z)| < \dfrac{\overline{r}_k}{1+\overline{r}_k}\rho_f$. An approximate zero for f relative to all $k > 0$ will be simply called an approximate zero.

Notice that the above concept is different from the one introduced in Definition 1.2 in Chapter 3. Since $\overline{r}_k$ increases in k, $\overline{r}_k/(1+\overline{r}_k) = 1-1/(1+\overline{r}_k)$ is also increasing in k and tends to $(3-\sqrt{8})/(1+3-\sqrt{3}) = 0.146446\cdots$. By some calculation, we have

$$
\frac{1}{12} < 0.085815 \leq \frac{\overline{r}_1}{1+\overline{r}_1} \leq \frac{\overline{r}_k}{1+\overline{r}_k} < \frac{1}{6}
$$

and for $k \geq 5$,

$$\frac{1}{7} < 0.14329 \leq \frac{\bar{r}_5}{1 + \bar{r}_5} \leq \frac{\bar{r}_k}{1 + \bar{r}_k} < \frac{1}{6}.$$

Therefore, if $|f(z)| < \rho_f/12$ then z is an approximate zero of f.

Suppose that $\rho_f = \min_{\theta, f(\theta)=0} |f(\theta)| > 0$ and ξ is a zero of f. Obviously, $f'(\xi) \neq 0$ since $\rho_f > 0$. Besides, f_ξ^{-1} may be uniquely analytically continued along any ray starting from 0 as long as the inverse image of this ray doesn't run into a critical point of f. Thus f_ξ^{-1} may be analytically continued to $\mathbf{C} - \bigcup_{j=1}^{k}(f(\theta_j), \infty)$, where $\theta_1, \cdots, \theta_k$ are all distinct critical points of f and $(f(\theta_j), \infty)$ represents a ray starting from $f(\theta_j)$. Denote by $S_{\xi, f}$ the domain $\mathbf{C} - \bigcup_{j=1}^{k}(f(\theta_j), \infty)$.

Let f be a polynomial of degree n with $n > 1$. Then k above satisfies $1 \leq k \leq n - 1$. We still use $f_\xi^{-1} : S_{\xi, f} \to \mathbf{C}$ to denote the above analytic continuation. It is obvious that the images of $f_{\xi_j}^{-1}$ are disjoint for all distinct zeros ξ_j of f since the inverse image of every the ray by $f_{\xi_j}^{-1}$ is a solution curve $z(t)$, $f(z(t)) = f(z(t_0))e^{-(t-t_0)}$ of the Newton differential equation $dz/dt = -f(z)/f'(z)$ in $\mathbf{R}^2 = \mathbf{C}$ which terminates at ξ_j

Consider now a general $z \in \mathbf{C}$. If f_z^{-1} can be analytically continued along the ray from $f(z)$ to 0 then there is some zero ξ of f such that $f_z^{-1} = f_\xi^{-1}$ in a neighborhood of the ray. We analytically continue $f_z^{-1} : S_{z,f} \to \mathbf{C}$ by setting $S_{z,f} = S_{\xi,f}$ and $f_z^{-1} = f_\xi^{-1}$.

Let $\rho_{f,z}$ be the convergent radius of $f_z^{-1} : S_{z,f} \to \mathbf{C}$ around 0. Obviously, $\rho_f = \min \rho_{f,z}$. Moreover, if $z' \in \text{Image}(f_z^{-1})$ then $S_{z',f} = S_{z,f}$ and $f_{z'}^{-1} = f_z^{-1}$.

Definition 3.2 If $|f(z)| < \dfrac{\bar{r}_k}{1 + \bar{r}_k}\rho_{f,z}$ then z is called a generalized approximate zero for f relative to k. If for every $k > 0$, z is a generalized approximate zero for f relative to k then z is called a generalized approximate zero of f.

For the further discussion of the approximate zeros and the generalized approximate zeros, we give a slight extension of Bieberbach-Koebe Theorem.

Theorem 3.3 (Extended Bieberbach-Koebe Theorem)   *Let f be a one-to-one analytic function on $\{z : |z| < r\}$. Then*

$$\text{Image}(f) \supset \{w : |w - f(0)| < |f'(0)|r/4\}.$$

Proof. Let $g(\tilde{z}) = \dfrac{f(r\tilde{z}) - f(0)}{f'(0)r}$ for $|\tilde{z}| < 1$. Then

$$g(0) = 0, \quad g'(0) = \frac{1}{f'(0)r} \cdot rf'(r\tilde{z})|_{z=0} = 1.$$

Thus g is Schlicht. By Bieberbach-Koebe Theorem in Section 2,

$$\text{Image}(g) \supset \{\tilde{w} : |\tilde{w}| < 1/4\}.$$

Hence,

$$\text{Image}(f) \supset \{w : |w - f(0)| < |f'(0)|r/4\}.\P$$

Then we have

Theorem 3.4 *Let $\xi_1, \cdots, \xi_k$ be all the distinct simple zeros of*

$$f(z) = z^n + a_{n-1}z^{n-1} + \cdots + a_n.$$

Then

$$\left\{ z : |z - \xi_i| < \frac{1}{4} \cdot \frac{\bar{r}_k}{1 + \bar{r}_k} \cdot \frac{\rho_{f,\xi_i}}{|f'(\xi_i)|} \right\}, \quad i = 1, \cdots, k$$

are disjoin one another and consist of only generalized approximate zeros of f.

Proof. By Theorem 3.3, the image by $f_{\xi_i}^{-1}$ of

$$\left\{ w : |w| < \frac{\bar{r}_k}{1 + \bar{r}_k} \rho_{f,\xi_i} \right\}$$

contains the disk

$$\left\{ z : |z - f_{\xi_i}^{-1}(0)| < \frac{1}{4} \cdot \frac{\bar{r}_k}{1 + \bar{r}_k} \rho_{f,\xi_i} |(f_{\xi_i}^{-1})'(0)| \right\}$$
$$= \left\{ z : |z - \xi_i| < \frac{1}{4} \cdot \frac{\bar{r}_k}{1 + \bar{r}_k} \frac{\rho_{f,\xi_i}}{|f'(\xi_i)|} \right\}.$$

Since z belonging to the disk implies $|f(z)| < \dfrac{\bar{r}_k}{1 + \bar{r}_k}\rho_{f,\xi_i}$, the disk consists of only generalized approximate zeros of f.

The first part of this theorem is directly from the result that the images of $f_{\xi_i}^{-1} : S_{\xi_i,f} \to \mathbf{C}$ are disjoint one another.$\P$

Next lemma is technical.

Lemma 3.5 *Let $p \in D_1 \subset \mathbf{R}^2$ and $0 < r < 1$. Then the area of $D_r(p) \cap D_1$ satisfies*

$$\text{Area}(D_r(p) \cap D_1) > \frac{r^2}{2}\sqrt{4 - r^2},$$

where $D_r(p) = \{z : |z - p| < r\}$.

Proof. It is easy to see that the y-coordinate of the intersecting point of the two circles $x^2 + y^2 = 1$ and $x^2 + (y - \theta)^2 = r^2$ is

$$y(\theta) = \frac{1 + \theta^2 - r^2}{2\theta} = \frac{1}{2}\left(\theta + \frac{1 - r^2}{\theta}\right).$$

Thus $y(1) = (2 - r^2)/2$.

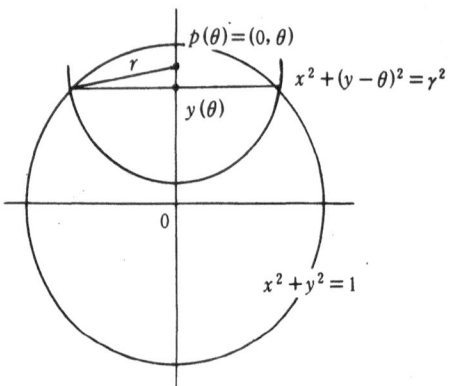

Figure 5.3

When $\sqrt{1 - r^2} < \theta \leq 1$, $y'(\theta) = \frac{1}{2}\left(1 - \frac{1 - r^2}{\theta^2}\right) > 0$. Thus $y(\theta)$ is increasing in θ. Moreover, since $1 - \sqrt{1 - r^2} < r$, it is clear that for $\sqrt{1 - r^2} < \theta_1 < \theta_2 \leq 1$,

$$D_r(p(\theta_1)) \cap D_1 \supset D_r(p(\theta_2)) \cap D_1.$$

Therefore, when $0 \leq \theta \leq \sqrt{1 - r^2}$,

$$\text{Area}(D_r(p(\theta)) \cap D_1) \geq \frac{\pi}{2}r^2.$$

When $\sqrt{1 - r^2} < \theta \leq 1$,

$$\text{Area}(D_r(p(\theta)) \cap D_1) \geq \text{Area}(D_r(p(1)) \cap D_1)$$

$$\geq\ 2\cdot\frac{1}{2}r\cdot\sqrt{\left(1-\left(\frac{2-r^2}{2}\right)^2\right)}$$

$$=\ \frac{r^2\sqrt{4-r^2}}{2}.$$

This proves the lemma. ¶

Based on Theorem 3.4 and Lemma 3.5, the following theorem provides an estimate of the area of approximate zeros.

Theorem 3.6 *Let*

$$P_n(R) = \{g \in \mathcal{P}_n \ : \ g(z) = z^n + b_{n-1}z^{n-1} + \cdots + b_1z + b_0,$$
$$|b_j| < R \ for \ j = 0, 1, \cdots, n-1\}.$$

If $f \in P_n(1)$ and $\rho_f > 0$ then

$$\text{Area}\left\{z \in \mathbf{C} : |z| < 1, |f(z)| < \frac{\overline{r}_k}{1+\overline{r}_k}\rho_f\right\} > 0.00173 \left(\frac{\rho_f}{n(n+1)}\right)^2,$$

where n is the degree of f.

Proof. Suppose $f(z) = z^n + a_{n-1}z^{n-1} + \cdots + a_1z + a_0$ with $|a_j| < 1$ for $j = 0, 1, \cdots, n-1$. Suppose that $\xi_1, \cdots, \xi_n$ are all the zeros of f. Obviously, $|\xi_1 \cdots \xi_n| = |a_0| < 1$. Thus there is a zero ξ of f such that $|\xi| < 1$. So

$$\begin{aligned}|f'(\xi)| &= |n\xi^{n-1} + (n-1)a_{n-1}\xi^{n-2} + \cdots + a_1| \\ &< n + (n-1) + \cdots + 1 \\ &= \frac{n(n+1)}{2}.\end{aligned}$$

Therefore,

$$\left|\frac{1}{f'(\xi)}\right| > \frac{2}{n(n+1)}.$$

By Theorem 3.4, we obtain that the open disks

$$D_\alpha(\xi) \supset D_\beta(\xi)$$

consist of the approximate zeros of f, where

$$\alpha = \frac{1}{4}\frac{\overline{r}_k}{1+\overline{r}_k}\frac{\rho_f}{|f'(\xi)|}, \quad \beta = \frac{1}{2}\frac{\overline{r}_k}{1+\overline{r}_k}\frac{\rho_f}{n(n+1)}.$$

By Lemma 3.5, we have

$$\text{Area}\left\{z \in \mathbf{C} : |z| < 1, |f(z)| < \frac{\overline{r}_k}{1 + \overline{r}_k}\rho_f\right\}$$

$$\geq \quad \text{Area}(D_\alpha(\xi) \bigcap D_1)$$

$$\geq \quad \text{Area}(D_\beta(\xi) \bigcap D_1)$$

$$> \quad \frac{\beta^2}{2}\sqrt{4 - \beta^2}.$$

Let $\theta_1, \cdots, \theta_{n-1}$ be the all zeros of $f'(z) = nz^{n-1} + (n-1)z^{n-2} + \cdots + a_1$. Since $|\theta_1 \cdots \theta_{n-1}| = |a_1|/n < 1/n \leq 1$, there must exist a critical point θ_* of f with $|\theta_*| < 1$. Thus

$$|f(\theta_*)| = |\theta_*^n + a_{n-1}\theta_n^{n-1} + \cdots + a_0| < n + 1.$$

Therefore,

$$\rho_f = \min_{\theta, f'(\theta)=0} |f(\theta)| < n + 1.$$

Notice that $\dfrac{1}{12} < \dfrac{\overline{r}_k}{1 + \overline{r}_k} < \dfrac{1}{6}$, we have

$$
\begin{aligned}
\frac{\beta^2}{2}\sqrt{4 - \beta^2} &= \frac{1}{2} \cdot \frac{\left(\frac{\overline{r}_k}{1 + \overline{r}_k}\rho_f\right)^2}{(2n(n+1))^2}\sqrt{4 - \left(\frac{\overline{r}_k}{1 + \overline{r}_k}\rho_f\right)^2 \Big/ (2n(n+1))^2} \\
&\geq \frac{1}{2}\frac{(\frac{1}{12})^2\rho_f^2}{4(n(n+1))^2}\sqrt{4 - \left(\frac{1}{6}(n+1)\right)^2 \Big/ 4n^2(n+1)^2} \\
&\geq \frac{1}{1152}\sqrt{4 - \frac{1}{144}}\left(\frac{\rho_f}{n(n+1)}\right)^2 \\
&> 0.00173\left(\frac{\rho_f}{n(n+1)}\right)^2.\P
\end{aligned}
$$

In the rest of this section we discuss the incremental Euler algorithms.

Lemma 3.7   Let $f(z) = z + a_2z^2 + a_3z^3 + \cdots$ be a one-to-one analytic function on D_{h_*}. Denote $r = |z|/h_*$ for $|z| < h_*$. Then

(1) $|f(z)| \leq |z|/(1-r)^2$;

(2)   For all $0 < r < r_k$,

$$|f(z) - T_kf(z)| \leq \frac{h_*(k+1)r^{k+1}}{(1-r)^2}.$$

Proof. Let $\tilde{f}(\tilde{z}) = f(h_*\tilde{z})/h_*$. Obviously, $\tilde{f}(0) = 0$ and $\tilde{f}'(0) = 1$. So f is Schlicht.

(1) By Koebe-Gronwall Theorem, we have

$$|\tilde{f}(\tilde{z})| \leq \frac{|\tilde{z}|}{(1 - |\tilde{z}|)^2}.$$

Denote $z = h_*\tilde{z}$. Then

$$|f(z)| = |f(h_*\tilde{z})| = h_*|\tilde{f}(\tilde{z})| \leq h_* \frac{|z/h_*|}{(1-r)^2} = \frac{|z|}{(1-r)^2}.$$

(2) From the proof of Theorem 2.9,

$$|\tilde{f}(\tilde{z}) - T_k\tilde{f}(z)| \leq \frac{(k+1)|\tilde{z}|^{k+1}}{(1 - |\tilde{z}|)^2}.$$

Substituting $z = h_*\tilde{z}$ with $|\tilde{z}| = r$ for z in the above inequality, we have

$$|f(z) - T_kf(z)| = h_*|\tilde{f}(\tilde{z}) - T_k\tilde{f}(\tilde{z})| \leq \frac{h_*(k+1)r^{k+1}}{(1-r)^2}.\P$$

Lemma 3.8 *Let $0 < h_* < h_1(f, z)$ and $h = rh_*$ with $0 < r < r_k$. Then*

$$T_k\sigma^{-1}(h) \in \sigma^{-1}(D_{h_*}).$$

Proof. Applying Lemma 3.3 to σ^{-1} defined on D_{h_*}, we obtain

$$|\sigma^{-1}(h)| \leq \frac{h}{(1-r)^2}$$

and

$$|\sigma^{-1}(h) - T_k\sigma^{-1}(h)| \leq \frac{h_*(k+1)r^{k+1}}{(1-r)^2}.$$

Thus

$$
\begin{aligned}
|T_k\sigma^{-1}(h)| &\leq |\sigma^{-1}(h)| + |T_k\sigma^{-1}(h) - \sigma^{-1}(h)| \\
&\leq \frac{h}{(1-r)^2} + \frac{h_*(k+1)r^{k+1}}{(1-r)^2} \\
&< \frac{h_*}{4}.
\end{aligned}
$$

The last inequality is equivalent to

$$(1 - r)^2 - 4r(1 + r^k(k + 1)) > 0.$$

But the latter is obviously true by the definition of r_k and $0 < r < r_k$. Then Theorem 3.3 gives $\sigma^{-1}(D_{h_*}) \supset D_{|(\sigma^{-1})'(0)|h_*/4} = D_{h_*/4}$. Thus

$$T_k \sigma^{-1}(h) \in \sigma^{-1}(D_{h_*}).\P$$

Now, we are ready to give the effect of one iteration of the incremental Euler algorithm.

Theorem 3.9   Let $0 < h_* \le h_1(f, z)$ and $h = rh_*$ with $0 < r < r_k$. Then for the iteration

$$z' = E_{k(h,f)}(z),$$

there exists a complex number h' with $|h'| < h_*$ such that

$$z' = f_z^{-1}((1 - h')f(z)).$$

Proof.   Theorem 1.9 gives

$$z' = z + FT_k(\sigma^{-1}(h))$$

while Lemma 3.8 guarantees

$$T_k(\sigma^{-1}(h)) \in \sigma^{-1}(D_{h_*}).$$

Thus there is a complex number h' with $|h'| < h_*$ such that

$$T_k(\sigma^{-1}(h)) = \sigma^{-1}(h').$$

Therefore,
$$z' = z + F\sigma^{-1}(h') = f_z^{-1}((1 - h')f(z)).\P$$

Furthermore, we have

Theorem 3.10   Let $z_* = f_z^{-1}(0)$, where $f_z^{-1} : S_{z,f} \to \mathbf{C}$. If we substitute $\rho_{f,z}$ for ρ_f then the result of Theorem 2.15 is still true.

Proof.   Theorem 3.9 says $z' = E_{k(h,f)}(z) = f_z^{-1}((1 - h')f(z))$ and $z_l = (E_k)'(z) \in f_z^{-1}(S_{z,f})$. With the argument similar to the proof of Theorem 2.15 we can obtain the result.¶

§4. E_k Iteration

This section concerns the incremental Euler algorithm E_k on a wedge shaped circular sector.

Definition 4.1   Let $f_z^{-1} : S_{z,f} \to \mathbf{C}$ be as in Section 3 with $f'(z) \neq 0$. For $0 < \alpha \leq \pi/2$ and $f(z) \neq 0$, define

$$w_{f,z,\alpha} = \left\{ w \in \mathbf{C} : |w| \leq 2|f(z)|, \left| \arg \frac{w}{f(z)} \right| < \alpha \right\}.$$

Let $\alpha_* = \sup\{\alpha | f_z^{-1}$ be analytic on $w_{f,z,\alpha}, 0 < \alpha \leq \pi/2\}$. Obviously, f_z^{-1} is analytic on the wedge w_{f,z,α_*}. Moreover, if f_z^{-1} is analytic on $w_{f,z,\alpha}$ then $w_{f,z,\alpha} \subset w_{f,z,\alpha_*}$. Denote $w_{f,z} = w_{f,z,\alpha_*}$ and $\theta_{f,z} = \alpha_*$. (cf. Fig. 5.4)

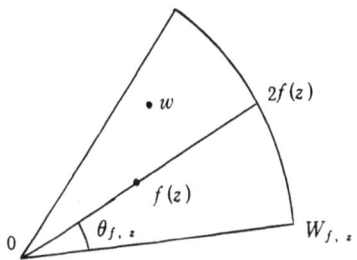

Figure 5.4

It is easy to see that $\theta_{f,z} > 0$ if and only if there is no $\lambda \in [0,2]$ such that $f'(\theta) = 0$ and $f(\theta) = \lambda f(z)$. The degenerate case is that there is a $\lambda \in [0,2]$ such that $f'(\theta) = 0$ and $f(\theta) = \lambda f(z)$. As a convention we use $\theta_{f,z} = 0$ to denote the degenerate case.

For the convenience of discussion, we introduce

Definition 4.2   Denote

$$K(h) = \frac{(k+1)^{\frac{k+1}{k}}}{k \bar{r}_k (1 - \bar{r}_k)^{\frac{1}{k}}}.$$

It is obvious that

$$\lim_{k \to \infty} K(k) = \lim_{k \to \infty} \frac{k+1}{k} \cdot \frac{1}{\bar{r}_k} \cdot \frac{(k+1)^{\frac{1}{k}}}{(1 - \bar{r}_k)^{\frac{1}{k}}} = \frac{1}{3 - \sqrt{8}} = 5.828 \cdots.$$

Lemma 4.3   *Let $c > 0, 0 < a < 1$ and $\alpha_k(r) = \beta_k(r)r^k$. Then the equation*

$$(k+1)c + 1 = \frac{1-h}{\alpha_k(h/\alpha)}$$

has a unique solution $h = h_0$ which satisfies $0 < h_0 < a\bar{r}_k$ and

$$h_0 \geq \frac{a(k+1)}{kK(k)(c+1)^{1/k}}.$$

Furthermore, for $0 < h \leq h_0$,

$$\frac{1}{1 - \alpha_k(h/a)} \leq 1 + \frac{1}{c(k+1)}.$$

Proof.   Since $\alpha_k(r) = \beta_k(r)r^k$, $\alpha_k(0) = 0$ and $\alpha_k(\bar{r}_k) = \beta_k(\bar{r}_k)\bar{r}_k^k = 1$, it is clear that

$$\lim_{h \to 0^+} \frac{1-h}{\alpha_k(h/a)} = +\infty$$

and

$$\lim_{h \to a\bar{r}_k} \frac{1-h}{\alpha_k(h/a)} = \frac{1 - a\bar{r}_k}{\alpha_k(\bar{r}_k)} = 1 - a\bar{r}_k < 1.$$

Moreover, since $\alpha_k(h/a)$ is strictly increasing in h, $(1-h)/\alpha_k(h/a)$ is strictly decreasing in h. Notice that $(k+1)c + 1 > 1$, we obtain that

$$(k+1)c + 1 = \frac{1-h}{\alpha_k(h/a)}$$

has a unique solution h_0 with $0 < h_0 < a\bar{r}_k$. For $0 < h \leq h_0$, $\alpha_k(h/a) \leq \alpha_k(h_*/a)$. By the definition of h_0, we have

$$\frac{1}{1 - \alpha_k(h/a)} \leq \frac{1}{1 - \alpha_k(h_0/a)}$$

$$= \frac{1}{1 - (1 - h_0)/((k+1)c + 1)}$$

$$\leq \frac{1}{1 - 1/((k+1)c + 1)}$$

$$= \frac{(k+1)c + 1}{(k+1)c}$$

$$= 1 + \frac{1}{c(k+1)}.$$

Now, we prove the first part of this lemma. Notice that $h_0 \leq a\bar{r}_k < \bar{r}_k$ $(0 < a < 1)$. By the definition of h_0, we have

$$\left(\frac{h_0}{a}\right)^k \beta_k\left(\frac{h_0}{a}\right) = \alpha_k\left(\frac{h_0}{a}\right) = \frac{1 - h_0}{c(k+1)+1} \geq \frac{1 - \bar{r}_k}{c(k+1)+1}.$$

Thus

$$
\begin{aligned}
\left(\frac{h_0}{a}\right)^k &\geq \frac{1 - \bar{r}_k}{c(k+1)+1} \cdot \frac{1}{\beta_k(h_0/a)} \\
&\geq \frac{1 - \bar{r}_k}{c(k+1)+1} \cdot \frac{1}{\beta_k(\bar{r}_k)} \\
&= \frac{1 - \bar{r}_k}{c(k+1)+1)} \cdot \frac{\bar{r}_k^k}{\beta_k(\bar{r}_k)\bar{r}_k^k} \\
&= \frac{1 - \bar{r}_k}{c(k+1)+1}\bar{r}_k^k.
\end{aligned}
$$

This implies

$$
\begin{aligned}
h_0 &\geq a\frac{\bar{r}_k(1 - \bar{r}_k)^{1/k}}{(c(k+1)+1)^{1/k}} \\
&\geq \frac{a\bar{r}_k(1 - \bar{r}_k)^{1/k}}{(k+1)^{1/k}(c+1)^{1/k}} \\
&= \frac{a(k+1)}{kK(k)(c+1)^{1/k}}. \quad \P
\end{aligned}
$$

Lemma 4.4 $h_1(f, z) \geq \sin\theta_{f,z}$.

Proof. If $\theta_{f,z} = \pi/2$ then the wedge $w_{f,z}$ is a semi-circle of radius $2|f(z)|$ while $w_{f,z} \supset D_{|f(z)|}(f(z))$. Thus (cf. Fig.5.5)

$$h_1(f, z) \geq \frac{|f(z)|}{|f(z)|} = 1 = \sin\frac{\pi}{2}.$$

If $\theta_{f,z} < \pi/2$ then any critical point θ of f must lie on the boundary of $w_{f,z}$ or outside of $w_{f,z}$. Referring to Fig. 5.6, we have

$$
\begin{aligned}
h_1(f, z) &= \min_{\theta, f'(\theta)=0} \frac{|f(z) - f(\theta)|}{|f(z)|} \\
&\geq \min_{\theta, f'(\theta)=0} \frac{t}{|f(z)|} \\
&= \sin\theta_{f,z}. \quad \P
\end{aligned}
$$

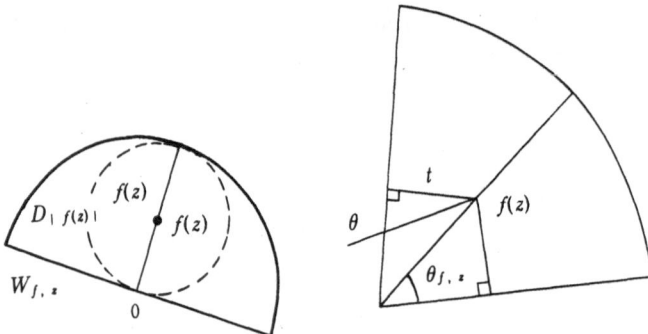

Figure 5.5   Figure 5.6

The proof of next lemma is straightforward.

Lemma 4.5 *If $0 \leq x \leq \dfrac{\pi}{2}$ and $0 \leq \alpha \leq 1$ then*

(1) $0 \leq \arctan x \leq x$;

(2) $\sin \alpha x \geq \alpha \sin x$. ¶

Next lemma shows the effect of one iteration of the incremental Euler algorithm.

Lemma 4.6 *Let $r = \dfrac{h}{h_1}, 0 < r < \bar{r}_k$ and $z' = E_{k(h,f)}(z)$. Then*

(1) $\left| \dfrac{f(z')}{f(z)} \right| \leq 1 - (1 - \alpha_k(r))h$;

(2) $\left| \arg \dfrac{f(z')}{f(z)} \right| < \dfrac{\alpha_k(r)h}{1 - (1 + \alpha_k(r))h}$.

Proof. (1) Since $|Q(h, f, z)| \leq \beta_k(r)$ and $\alpha_k(r) = \beta_k(r)r^k$, by Theorem 2.10, we have

$$
\begin{aligned}
\left| \frac{f(z')}{f(z)} \right| &= |1 - h + Q(h, f, z)r^k h| \\
&\leq 1 - h + \alpha_k(r)h \\
&= 1 - (1 - \alpha_k(r))h.
\end{aligned}
$$

(2) Referring to Fig. 5.7, by Theorems 2.11 and 4.5(1), we have

$$
\left| \arg \frac{f(z')}{f(z)} \right| = \arctan \frac{|\mathrm{Im}Q(h, f, z)h^{k+1}/h_1^k|}{|1 - h + \mathrm{Re}Q(h, f, z)h^{k+1}/h_1^k|}
$$

$$\leq \left| \frac{\operatorname{Im} Q(h,f,z) r^k h}{1 - h + \operatorname{Re} Q(h,f,z) r^k h} \right|$$

$$< \frac{\beta_k(r) r^k h}{1 - h - \beta_k(r) r^k h}$$

$$= \frac{\alpha_k(r) h}{1 - h - \alpha_k(r) h}$$

$$= \frac{\alpha_k(r) h}{1 - (1 + \alpha_k(r)) h} \quad \P$$

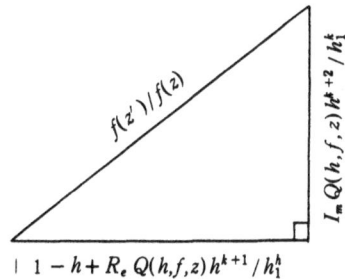

Figure 5.7

Lemma 4.7 *Let* $h_* = \dfrac{k}{k+1} \sin \theta_{f,z_0} > 0$ *(so* $f(z_0) \neq 0$*),* $\delta = h/h_*$*,* $0 < \delta < \bar{r}_k$ *(i.e.* $0 < h < h_* \bar{r}_k$*), and* $z_l = E^l_{k(h,f)}(z_0)$*. Then for* l *satisfying*

$$0 \leq l \leq \frac{1 - (1 + \alpha_k(\delta) h)}{(k+1) \alpha_k(\delta) h} \theta_{f,z_0},$$

we have

$$\theta_{f,z_l} \geq \frac{k}{k+1} \theta_{f,z_0}$$

and

$$|f(z_{l+1})| < (1 - (1 - \alpha_k(\delta)) h)^{l+1} |f(z_0)|.$$

Proof. The proof goes by induction on l. When $l = 0$, it is easy to see that

$$\theta_{f,z_0} \geq \frac{k}{k+1} \theta_{f,z_0}.$$

Moreover, by Lemmas 4.4 and 4.5(2),

$$h_1(f, z_0) \geq \sin \theta_{f,z_0} \geq \sin \frac{k}{k+1} \theta_{f,z_0} > \frac{k}{k+1} \sin \theta_{f,z_0}$$

and
$$r = r(f, z_0) = \frac{h}{h_1(f, z_0)} < \frac{h}{\frac{k}{k+1} \sin \theta_{f, z_0}} = \frac{h}{h_*} = \delta < \bar{r}_k.$$

Then by Lemma 4.6(1),

$$\left| \frac{f(z_1)}{f(z_0)} \right| \leq 1 - (1 - \alpha_k(r))h < 1 - (1 - \alpha_k(\delta))h,$$

that is,

$$|f(z_1)| < (1 - (1 - \alpha_*(\delta))h)|f(z_0)|.$$

Now, suppose that the lemma is true for $l - 1$, that is, when

$$0 \leq l - 1 \leq \frac{1 - (1 + \alpha_k(\delta)h)}{(k + 1)\alpha_k(\delta)h} \theta_{f, z_0},$$

we have

$$\theta_{f, z_{l-1}} \geq \frac{k}{k + 1} \theta_{f, z_0}$$

and

$$|f(z_{l-1})| < (1 - (1 - \alpha_k(\delta))h)^{l-1} |f(z_0)|.$$

Similarly, by Lemmas 4.4 and 4.5(2)

$$h_1(f, z_{l-1}) \geq \sin \theta_{f, z_{l-1}} \geq \sin \frac{k}{k + 1} \theta_{f, z_0} > \frac{k}{k + 1} \sin \theta_{f, z_0},$$

therefore,

$$r = r(f, z_{l-1}) = \frac{h}{h_1(f, z_{l-1})} < \frac{h}{\frac{k}{k+1} \sin \theta_{f, z_0}} = \frac{h}{h_*} = \delta < \bar{r}_k,$$

$$\left| \arg \frac{f(z_l)}{f(z_{l-1})} \right| < \frac{\alpha_k(r)h}{1 - (1 + \alpha_k(\delta))h} < \frac{\alpha_k(\delta)h}{1 - (1 + \alpha_k(\delta))h},$$

and

$$\left| \frac{f(z_l)}{f(z_{l-1})} \right| \leq 1 - (1 - \alpha_k(r))h < 1 - (1 - \alpha_k(\delta))h,$$

$$|f(z_l)| < (1 - (1 - \alpha_k(\delta))h)|f(z_{l-1})| \leq (1 - (1 - \alpha_k(\delta))h)^l |f(z_0)|.$$

We have shown that $h_* = \frac{k}{k + 1} \sin \theta_{f, z_0}$, $0 < h_* < h_1(f, z_{l-1})$, $h = \delta h_*$ and $0 < \delta < \bar{r}_k$. By Theorem 3.9, there is a complex number h' with $|h'| < h_*$ such that $z_l = f_{z_{l-1}}^{-1}((1 - h')f(z_{l-1}))$. Since (cf. Fig. 5.8)

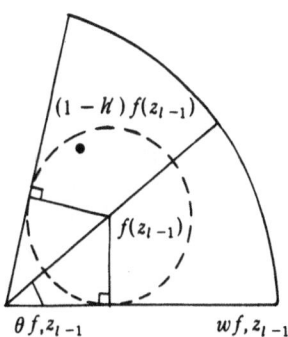

$(1 - h') f(z_{l-1})$

$f(z_{l-1})$

$\theta f, z_{l-1}$ $w f, z_{l-1}$

Figure 5.8

$$
\begin{aligned}
|(1 - h') f(z_{l-1}) - f(z_{l-1})| &= |h' f(z_{l-1})| \\
&\leq h_* |f(z_{l-1})| \\
&= \frac{k}{k+1} \sin \theta_{f,z_0} \cdot |f(z_{l-1})| \\
&< \sin \theta_{f,z_{l-1}} \cdot |f(z_{l-1})|,
\end{aligned}
$$

so

$$
(1 - h') f(z_{l-1}) \in w_{f,z_{l-1}} \text{ and } z_l \in \text{Image}(f_{z_{l-1}}^{-1}),
$$

where $f_{z_{l-1}}^{-1} : S_{z_{l-1},f} \to \mathbf{C}$. Obviously, $S_{z_l,f} = S_{z_{l-1},f}$.

The given condition

$$
0 \leq l \leq \frac{1 - (1 + \alpha_k(\delta)h)}{(k+1)\alpha_k(\delta)h} \theta_{f,z_0}
$$

is equivalent to

$$
\frac{l\alpha_k(\delta)h}{1 - (1 + \alpha_k(\delta)h)} \leq \frac{1}{k+1} \theta_{f,z_0}.
$$

Under this condition, we have

$$
\begin{aligned}
\theta_{f,z_l} &\geq \theta_{f,z_{l-1}} - \left| \arg \frac{f(z_l)}{f(z_{l-1})} \right| \\
&\geq \theta_{f,z_{l-1}} - \frac{\alpha_k(\delta)h}{1 - (1 + \alpha_k(\delta))h} \\
&\geq \cdots \\
&\geq \theta_{f,z_0} - \frac{l\alpha_k(\delta)h}{1 - (1 + \alpha_k(\delta))h} \\
&\geq \theta_{f,z_0} - \frac{1}{k+1} \theta_{f,z_0} = \frac{k}{k+1} \theta_{f,z_0}.
\end{aligned}
$$

Finally, with $\theta_{f,z_l} \geq \dfrac{k}{k+1}\theta_{f,z_0}$ and an argument similar to above, we obtain

$$|f(z_{l+1})| \leq (1 - (1 - \alpha_k(\delta))h)^{l+1}|f(z_0)|. \P$$

Now, we are ready to give the main result.

Theorem 4.8 *Given a polynomial f, a complex number z_0 and a real positive number L satisfying $|f(z_0)| > L$ and $\theta_{f,z_0} > 0$, let $c = \dfrac{1}{\theta_{f,z_0}} \ln \dfrac{|f(z_0)|}{L}$, then there is*

$$h_0 \geq \frac{\sin \theta_{f,z_0}}{K(k)(c+1)^{1/k}}$$

with the following property: for each h with $0 < h \leq h_0$, take

$$l = \left\lceil \frac{1}{h} \cdot \frac{1}{1 - \alpha_k(h/a)} \ln \frac{|f(z_0)|}{L} \right\rceil, \quad a = \frac{k}{k+1} \sin \theta_{f,z_0},$$

then $|f(z_l)| < L$, where $z_l = E_{k(h,f)}^l(z_0)$.

Proof. Applying Lemma 4.3 with

$$a = \frac{k}{k+1} \sin \theta_{f,z_0} \text{ and } c = \frac{1}{\theta_{f,z_0}} \ln \frac{|f(z_0)|}{L},$$

we obtain a unique solution $h = h_0$ of

$$(k+1)c + 1 = \frac{1-h}{\alpha_k(h/a)},$$

where h_0 satisfies

$$h_0 \geq \frac{a(k+1)}{kK(k)(c+1)^{1/k}} = \frac{\left(\dfrac{k}{k+1} \sin \theta_{f,z_0}\right)(k+1)}{kK(k)(c+1)^{1/k}} = \frac{\sin \theta_{f,z_0}}{K(k)(c+1)^{1/k}}.$$

Now, we will prove that $|f(z_l)| < L$ for $0 < h \leq h_0$ and

$$l = \left\lceil \frac{1}{h} \cdot \frac{1}{1 - \alpha_k(h/a)} \ln \frac{|f(z_0)|}{L} \right\rceil.$$

In fact, when $0 \leq h \leq h_0$,

$$\frac{1-h}{\alpha_k(\delta)} = \frac{1-h}{\alpha_k(h/a)} \geq \frac{1-h_0}{\alpha_k(h_0/a)} = (k+1)c + 1,$$

that is,

$$\frac{1-h}{\alpha_k(\delta)} - 1 \geq (k+1)c.$$

This implies

$$\frac{(1-\alpha_k(\delta))(1-(1+\alpha_k(\delta))h)}{\alpha_k(\delta)}$$

$$= \frac{1-h-\alpha_k(\delta)h - \alpha_k(\delta) + \alpha_k(\delta)h + \alpha_k(\delta)^2h}{\alpha_k(\delta)}$$

$$\geq (k+1)c,$$

$$\frac{1-(1+\alpha_k(\delta))h}{(k+1)\alpha_k(\delta)} \geq \frac{c}{1-\alpha_k(\delta)} = \frac{\ln\dfrac{|f(z_0)|}{L}}{\theta_{f,z_0}(1-\alpha_k(\delta))}.$$

Thus

$$\frac{(1-(1+\alpha_k(\delta))h)\theta_{f,z_0}}{(k+1)\alpha_k(\delta)h} \geq \frac{\ln\dfrac{|f(z_0)|}{L}}{(1-\alpha_k(\delta))h}.$$

So there is a positive integer l such that

$$(*)\begin{cases} \dfrac{(1-(1+\alpha_k(\delta))h)\theta_{f,z_0}}{(k+1)\alpha_k(\delta)h} \geq l-1, \\[4mm] l \geq \dfrac{\ln\dfrac{|f(z_0)|}{L}}{(1-\alpha_k(\delta))h}. \end{cases}$$

In particular, we can take

$$l = \left\lceil \frac{1}{h} \cdot \frac{1}{1-\alpha_k(h/a)} \ln\frac{|f(z_0)|}{L} \right\rceil.$$

By the second inequality of $(*)$ and $|\ln(1-u)| \geq u$ for $0 \leq u < 1$,

$$l \geq \frac{\ln\dfrac{|f(z_0)|}{L}}{(1-\alpha_k(\delta))h} \geq \frac{\ln\dfrac{|f(z_0)|}{L}}{|\ln(1-(1-\alpha_k(\delta))h)|}.$$

Thus

$$l\ln(1-(1-\alpha_k(\delta))h) \leq \ln\frac{|f(z_0)|}{L}$$

and so

$$(1-(1-\alpha_k(\delta))h)^l|f(z_0)| \leq L.$$

Notice that the first inequality of $(*)$ is just the upper bound for $l-1$ in Lemma 4.7, thus Lemma 4.7 gives

$$|f(z_l)| < (1 - (1 - \alpha_k(\delta))h)^l |f(z_0)| \le L. \P$$

Theorem 4.9 *Under the hypotheses of Theorem 4.8, for any* $h \in (0, h_0)$ *there is* l *satisfying*

$$l \le \frac{1}{h} \left[\ln \frac{|f(z_0)|}{L} + \frac{\theta_{f,z_0}}{k+1} \right]$$

such that

$$|f(z_l)| < L,$$

where $z_l = E^l_{k(h,f)}(z_0)$.

   Proof. By Lemma 4.3,

$$\frac{1}{h} \cdot \frac{1}{1 - \alpha_k(h/a)} \ln \frac{|f(z_0)|}{L} \; \le \; \frac{1}{h} \left(1 + \frac{1}{c(k+1)} \right) \ln \frac{|f(z_0)|}{L}$$

$$= \frac{1}{h} \left(1 + \frac{\theta_{f,z_0}}{c(k+1) \ln \frac{|f(z_0)|}{L}} \right) \ln \frac{|f(z_0)|}{L}$$

$$= \frac{1}{h} \left[\ln \frac{|f(z_0)|}{L} + \frac{\theta_{f,z_0}}{k+1} \right].$$

Then by Theorem 4.8, $|f(z_l)| < L$ for l satisfying

$$l = \left[\frac{1}{h} \cdot \frac{1}{1 - \alpha_k(h/a)} \ln \frac{|f(z_0)|}{L} \right] \le \frac{1}{h} \left[\ln \frac{|f(z_0)|}{L} + \frac{\theta_{f,z_0}}{k+1} \right]. \P$$

   Applying the above theorem to find approximate zeros, we have

Theorem 4.10 *Let* $0 < L \le \dfrac{\bar{r}_k}{1 + \bar{r}_k} \rho_f$. *Under the hypotheses of Theorem 4.8, we have*

$$|f(z_l)| < L \le \frac{\bar{r}_k}{1 + \bar{r}_k} \rho_f$$

for any $h \in (0, h_0)$ *and some* l *satisfying*

$$l \le \frac{1}{h} \left[\ln \frac{|f(z_0)|}{L} + \frac{\theta_{f,z_0}}{k+1} \right],$$

that is, z_l is an approximate zero of f. ¶

Remark 4.11   If an approximate zero z_l of f is located as in Theorem 4.10, then according to Theorem 2.15, the iteration

$$z_{l+m} = E_{k(1,f)}^m(z_l), \quad m = 1, 2, \cdots$$

is well defined and the sequence $\{z_{l+m}\}_{m=1}^\infty$ converges to some zero of f.

§5.   Cost Theory of E_k as An Euler's Algorithm

We have obtained the conditions for reaching approximate zeros. Now we can discuss the cost for locating approximate zeros.

Lemma 5.1   *Let $S_R^1 = \{z \in \mathbf{C} : |z| = R\}$ and $f(z) = z^n + a_{n-1}z^{n-1} + \cdots + a_0$ be a complex polynomial. If S_R^1 contains all the zeros of f in its interior, then for any $z \in S_R^1$,*

$$\mathrm{Re}\left(-\bar{z}\frac{f(z)}{f'(z)}\right) < 0.$$

Moreover, the solution of the Newton differential equation $\dfrac{dz}{dt} = -\dfrac{f(z)}{f'(z)}$ is transversal to S_R^1 and points inward.

Proof.   Let $\xi_1, \cdots, \xi_n$ be the all zeros of f. Then for $z \in S_R^1$,

$$
\begin{aligned}
\mathrm{Re}(\bar{z}\xi_j) &= \mathrm{Re}((x-iy)(u_j+iv_j)) = xu_j + yv_j \\
&\le \sqrt{(x^2+y^2)(u_j^2+v_j^2)} < x^2 + y^2 = \bar{z}z.
\end{aligned}
$$

This means that the ξ_j are in the half plane separated by the tangent space to the circle at z and containing the origin (cf. Fig. 5.9).

It follows that

$$\mathrm{Re}(\bar{z}(z - \xi_j)) > 0,$$

$$\mathrm{Re}\left(\frac{z}{z-\xi_j}\right) = \mathrm{Re}\frac{\bar{z}z}{\bar{z}(z-\xi_j)} > 0,$$

$$\sum_{j=1}^n \mathrm{Re}\left(\frac{z}{z-\xi_j}\right) > 0,$$

$$\operatorname{Re}\left(z\frac{f'(z)}{f(z)}\right) = \operatorname{Re}\left(z\frac{\sum_{j=1}^{n}(z-\xi_1)\cdots(\widehat{z-\xi_j})\cdots(z-\xi_n)}{\prod_{j=1}^{n}(z-\xi_j)}\right)$$

$$= \operatorname{Re}\left(\sum_{j=1}^{n}\frac{z}{z-\xi_j}\right)$$

$$= \sum_{j=1}^{n}\operatorname{Re}\left(\frac{z}{z-\xi_j}\right) > 0.$$

So $f'(z) \neq 0$ and

$$\operatorname{Re}\left(\overline{z}\frac{f(z)}{f'(z)}\right) = \operatorname{Re}\left(z\frac{\overline{f(z)}}{f'(z)}\right)\frac{1}{\frac{f(z)}{f'(z)}\left(\frac{\overline{f(z)}}{f'(z)}\right)}$$

$$= \frac{1}{\frac{f(z)}{f'(z)}\left(\frac{\overline{f(z)}}{f'(z)}\right)}\operatorname{Re}\left(z\frac{f(z)}{f'(z)}\right) > 0.$$

Thus the inner product

$$< z, -\frac{f(z)}{f'(z)} >= \operatorname{Re}\left(-\overline{z}\frac{f(z)}{f'(z)}\right) < 0.$$

This inequality implies that the solution of $\dfrac{dz}{dt} = -\dfrac{f(z)}{f'(z)}$ is transversal to S_R^1 and is pointing inward. ¶

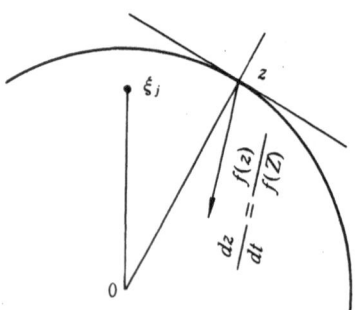

Figure 5.9

Lemma 5.1 gives the properties of f at a single point $z \in S_R^1$. Next lemma shows the relationship between the values $f(z_1)$ and $f(z_2)$ for any $z_1, z_2 \in S_R^1$.

Lemma 5.2 *Suppose that $f \in P_n(1)$ and $R \geq 2$. Let $z_1, z_2 \in S_R^1$ satisfy $|\arg(z_1/z_2)| \leq \beta < \pi/n$. Then*

$$n\beta - 2\arcsin\frac{1}{R-1} \leq \left|\arg\frac{f(z_1)}{f(z_2)}\right| \leq n\beta + 2\arcsin\frac{1}{R-1},$$

that is,

$$\left|n\beta - \left|\arg\frac{f(z_1)}{f(z_2)}\right|\right| \leq 2\arcsin\frac{1}{R-1}.$$

Proof. Since

$$\left|\frac{f(z)}{z^n} - 1\right| = \left|\frac{a_{n-1}z^{n-1} + \cdots + a_0}{z^n}\right|$$

$$\leq \frac{\sum_{j=0}^{n-1} R^j}{R^n} = \frac{(R^n - 1)/(R-1)}{R^n}$$

$$\leq \left(1 - \frac{1}{R^n}\right)\frac{1}{R-1} < \frac{1}{R-1},$$

so

$$\frac{f(z)}{z^n} \in D_{1/(R-1)}(1).$$

This implies $|\arg(f(z)/z^n)| < \arcsin(1/(R-1))$. With

$$\frac{f(z_1)}{f(z_2)} = \left(\frac{z_1}{z_2}\right)^n \frac{f(z_1)/z_1^n}{f(z_2)/z_2^n}$$

we have

$$n\beta - 2\arcsin\frac{1}{R-1} \leq \left|\arg\frac{f(z_1)}{f(z_2)}\right| \leq n\beta + 2\arcsin\frac{1}{R-1}.¶$$

Now, we turn to probabilistic discussions.

Lemma 5.3 *Let θ be a critical point of f. Denote by Θ_θ the branch of $f^{-1}(\lambda f(\theta))$ containing θ. Let $\Theta = \bigcup_{\theta, f'(\theta)=0} \Theta_\theta$. If $R \geq 2$ and $f \in P_n(1)$, then $\Theta \cap S_R^1$ is a set of at most $2(n-1)$ points.*

Proof. Recall that $\phi_t(z)$, the solution of the differential equation $\frac{dz}{dt} = -\frac{f(z)}{f'(z)}$ with $\phi_0(z) = z$, satisfies $f(\phi_t(z)) = e^{-t}f(z)$. By Lemma 5.1, for any fixed θ_j, $\Theta_{\theta_j} \cap S_R^1$ consists of at most $k_j + 1$ points, where $\theta_1, \cdots, \theta_m$ are the all distinct critical points of f and k_j is the multiplicity of θ_j as a zero of f'. Thus $\sum_{j=1}^m k_j = n - 1$. This proves that $\Theta \cap S_R^1$ has at most $\sum_{j=1}^m (k_j + 1) = \sum_{j=1}^m k_j + (n-1) = 2(n-1)$ points. ¶

Remark 5.4   Since $R \geq 2 > \max |a_j| + 1$, it is easy to show that all zeros of f lie in the interior of S_R^1.

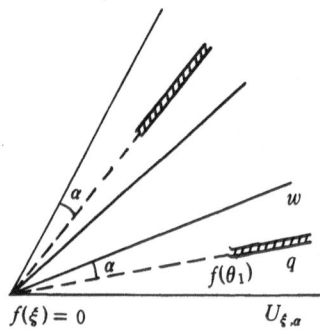

Figure 5.10

Lemma 5.5   *Let ξ be a zero of f and $f_\xi^{-1} : S_{\xi,f} \to \mathbf{C}$ be as in Section 3, where $S_{\xi,f} = \mathbf{C} - \bigcup_{j=1}^k (f(\theta_j), \infty)$. Given $\alpha > 0$, denote*

$$U_{\xi,\alpha} = \Big\{ w \in S_{\xi,f} \;:\; \text{there is some } q \in \bigcup_{j=1}^k (f(\theta_j), \infty)$$
$$\text{such that } |\arg \frac{w}{q}| < \alpha \Big\}.$$

When $\alpha < \pi/2$ and $z \notin f_\xi^{-1}(U_{\xi,\alpha})$ for any zero ξ of f, we have $\theta_{f,z} \geq \alpha$. Moreover, when $\alpha < \pi/2$ and $R \geq 1 + \sqrt{2}$, denote

$$N(\alpha) = S_R^1 \bigcap \bigcup_{\xi, f(\xi)=0} f_\xi^{-1}(U_{\xi,\alpha}),$$

then the measure of $N(\alpha)$ satisfies

$$\mathrm{vol}N(\alpha) < \frac{2(n-1)}{\pi n} \left(\alpha + 2 \arcsin \frac{1}{R-1} \right),$$

where vol is the normalized Lebesgue measure in S_R^1, i.e., $\mathrm{vol}(S_R^1) = 1$.

Proof.   First, we prove that the first part of the lemma is true. Otherwise, if $\theta_{f,z} < \alpha$ then there exists some ray containing $f(\theta)$ on the wedge $w_{f,z,\theta_{f,z}}$ with $f'(\theta) = 0$. But f_ξ^{-1} is not analytic at $f(\theta)$. Thus $f(z) \in U_{\xi,\alpha}$ and $z \in f_\xi^{-1}(U_{\xi,\alpha})$. This contradiction gives the result.

Now, we prove the second part. Keep in mind that f_ξ^{-1} have disjoint images for all distinct zeros of f. Lemma 5.3 shows that $N(\alpha)$ is contained in the set consisted of at most $2(n-1)$ wedges with center z_0 and angles $(\alpha + 2\arcsin(1/(R-1)))/n$, where $z_0 \in \Theta \cap S_R^1$.

Let

$$N(\alpha, z_0) = \left\{ z : z \in S_R^1, \left| \arg \frac{f(z)}{f(z_0)} \right| < \alpha \right\}.$$

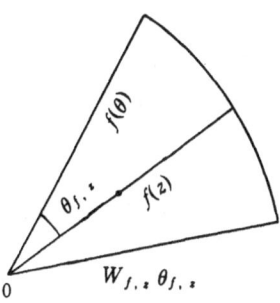

Figure 5.11

Since

$$\frac{f(z)}{f(z_0)} = \left(\frac{z}{z_0}\right)^n \frac{f(z)/z^n}{f(z_0)/z_0^n}$$

and

$$\left(\frac{z}{z_0}\right)^n = \frac{f(z)}{f(z_0)} \frac{f(z_0)/z_0^n}{f(z)/z^n},$$

we have $|\arg(z/z_0)| < \pi/n$. In fact, if $|\arg(z/z_0)| \geq \pi/n$ then there exists a $z_1 \in N(\alpha, z_0)$ such that $|\arg(z/z_0)| = \pi/n$. Thus

$$\pi = \left| \arg \frac{z_1^n}{z_0^n} \right| \leq \left| \arg \frac{f(z_1)}{f(z_0)} \right| + \left| \arg \frac{f(z_0)}{z_0^n} \right| + \left| \arg \frac{f(z_1)}{z_1^n} \right|$$

$$< \alpha + 2\arcsin \frac{1}{R-1} < \frac{\pi}{2} + \frac{\pi}{2} = \pi,$$

and this is impossible.

By Lemma 5.2,

$$n \left| \arg \frac{z}{z_0} \right| \leq \left| \arg \frac{f(z)}{f(z_0)} \right| + 2\arcsin \frac{1}{R-1} < \alpha + 2\arcsin \frac{1}{R-1}$$

and so

$$\left| \arg \frac{z}{z_0} \right| < \frac{1}{n} \left(\alpha + 2\arcsin \frac{1}{R-1} \right).$$

Thus we have

$$\text{vol}N(\alpha) < \sum_{z_0 \in \Theta \cap S_R^1} \text{vol}N(\alpha, z_0)$$

$$< \frac{4(n-1)}{2\pi} \frac{1}{n} \left(\alpha + 2 \arcsin \frac{1}{R-1} \right),$$

$$= \frac{2(n-1)}{\pi n} \left(\alpha + 2 \arcsin \frac{1}{R-1} \right). \P$$

Theorem 5.6 $\text{vol}\{f \in P_n(1) | \rho_f < \sigma\} \leq (n-1)\sigma^2$, where vol is the normalized Lebesgue measure in $P_n(1)$. $\P$

This theorem is a variant of Theorem 4.3.2.

Theorem 5.7 Let $0 < \alpha < \pi/2$, $R \geq 1 + \sqrt{2}$ and $f \in P_n(1)$. Then

$$\text{vol}\{z \in S_R^1 : \theta_{f,z} < \alpha\} \leq \frac{2(n-1)}{\pi n} \left(\alpha + 2 \arcsin \frac{1}{R-1} \right).$$

Proof. From the proof of Lemma 5.5, we see that $\theta_{f,z} < \alpha$ implies $z \in f_\xi^{-1}(U_{\xi,\alpha})$. Since $z \in S_R^1$,

$$z \in S_R^1 \cap \bigcup_{\xi, f(\xi)=0} f_\xi^{-1}(U_{\xi,\alpha}) = N(\alpha),$$

that is,

$$\{z \in S_R^1 : \theta_{f,z} < \alpha\} \subset N(\alpha).$$

Thus

$$\text{vol}\{z \in S_R^1 : \theta_{f,z} < \alpha\} \leq \text{vol}N(\alpha) \leq \frac{2(n-1)}{\pi n} \left(\alpha + 2 \arcsin \frac{1}{R-1} \right). \P$$

Theorem 5.8 *Denote*

$$Y_{\sigma,\alpha,R} = \{(z, f) \in S_R^1 \times P_n(1) : \rho_f < \sigma \text{ or } \theta_{f,z} < \alpha\}.$$

If $0 < \alpha < \pi/2$ *and* $R \geq 1 + \sqrt{2}$ *then*

$$\text{vol}Y_{\sigma,\alpha,R} \leq (n-1)\sigma^2 + \frac{2(n-1)}{\pi n} \left(\alpha + 2 \arcsin \frac{1}{R-1} \right).$$

Proof. From Theorems 5.6 and 5.7 we directly obtain

$$\begin{aligned}
\text{vol}Y_{\sigma,\alpha,R} &\leq \text{vol}\{(z, f) \in S_R^1 \times P_n(1) : \rho_f < \sigma\} \\
&\quad + \text{vol}\{(z, f) \in S_R^1 \times P_n(1) : \theta_{f,z} < \alpha\} \\
&\leq (n-1)\sigma^2 + \frac{2(n-1)}{\pi n} \left(\alpha + 2 \arcsin \frac{1}{R-1} \right). \P
\end{aligned}$$

Finally, we give the main result.

Theorem 5.9 *Given $k > 0$, $n > 0$ and $0 < \mu < 1$, there exist R, h and $M > 0$, $N > 0$ satisfying following condition: if $(z_0, f) \in S_R^1 \times P_n(1)$ then there is some l,*

$$l \leq M \left[\frac{n(|\ln \mu| + N)}{\mu} \right]^{\frac{k+1}{k}},$$

such that $z_l = E_{k(h,f)}^l(z_0)$ is an approximate zero of f with probability at least $1 - \mu$. Here the measure concerned is the normalized Lebesgue measure on the product space $S_R^1 \times P_n(1)$.

Proof. Suppose that k, n and μ are given. Then the system of equations

$$\begin{cases} (n-1)\sigma^2 = \dfrac{\mu}{10} \\ \dfrac{4(n-1)}{\pi n} \arcsin \dfrac{1}{R-1} = \dfrac{\mu}{10} \\ \dfrac{2(n-1)}{\pi n} \alpha = \dfrac{4\mu}{5} \end{cases}$$

in variables σ, R, α has a solution

$$\begin{cases} \sigma = \left(\dfrac{\mu}{10(n-1)} \right)^{1/2} \\ R = 1 + \left(\sin \dfrac{\mu\pi}{40} \left(\dfrac{n}{n-1} \right) \right)^{-1} \\ \alpha = \dfrac{2\pi\mu}{5} \left(\dfrac{n}{n-1} \right). \end{cases}$$

By Theorem 5.8, we obtain

$$\text{vol} Y_{\sigma,\alpha,R} \leq \frac{\mu}{10} + \frac{4\mu}{5} + \frac{\mu}{10} = \mu$$

and

$$\text{vol}(S_R^1 \times P_n(1) - Y_{\sigma,\alpha,R}) \geq 1 - \mu.$$

From the definition of $Y_{\sigma,\alpha,R}$, we know that $(z_0, f) \in S_R^1 \times P_n(1) - Y_{\sigma,\alpha,R}$ implies $\rho_f \geq \sigma$ and $\theta_{f,z_0} \geq \alpha$. Moreover, it is obvious that

$$|f(z_0)| = \left| \sum_{j=0}^n a_j z_0^j \right| \leq \sum_{j=0}^n R^j = \frac{R^{n+1} - 1}{R - 1}.$$

Thus, applying Theorem 4.10 with σ substituting for ρ_f, α for θ_{f,z_0} and $\dfrac{R^{n+1} - 1}{R - 1}$ for $|f(z_0)|$, and with $L = \dfrac{\bar{r}_k}{1 + \bar{r}_k}\sigma$, we obtain

$$c = \frac{1}{\theta_{f,z_0}} \ln \frac{|f(z_0)|}{L} \leq \frac{1}{\alpha} \ln \frac{(R^{n+1} - 1)/(R - 1)}{\dfrac{\bar{r}_k}{1 + \bar{r}_k}\sigma},$$

$$h_0 \geq \frac{\sin \theta_{f,z_0}}{K(k)(c + 1)^{1/k}}$$

$$\geq \frac{\sin \alpha}{K(k)\left[1 + \dfrac{1}{\alpha} \ln \dfrac{(R^{n+1} - 1)/(R - 1)}{\dfrac{\bar{r}_k}{1 + \bar{r}_k}\sigma}\right]^{1/k}}$$

$$= h$$

$$> 0,$$

$$l \leq \frac{1}{h}\left[\ln \frac{|f(z_0)|}{L} + \frac{\theta_{f,z_0}}{k + 1}\right]$$

$$\leq \frac{K(k)\left[1 + \dfrac{1}{\alpha} \ln \left(\dfrac{R^{n+1} - 1}{R - 1} \Big/ \dfrac{\bar{r}_k}{1 + \bar{r}_k}\sigma\right)\right]^{1/k}}{\sin \alpha}$$

$$\left[\ln \left(\frac{R^{n+1} - 1}{R - 1} \Big/ \frac{\bar{r}_k}{1 + \bar{r}_k}\sigma\right) + \frac{\pi/2}{k + 1}\right]$$

$$= \frac{K(k)\left[1 + \dfrac{1}{\alpha} \ln R^n + \dfrac{1}{\alpha} \ln \dfrac{R - 1/R^n}{R - 1} - \dfrac{1}{\alpha} \ln \dfrac{\bar{r}_k}{1 + \bar{r}_k}\sigma\right]^{1/k}}{\alpha \cdot \dfrac{\sin \alpha}{\alpha}}$$

$$\times \left[\ln R^n + \ln \frac{R - 1/R^n}{R - 1} - \ln \frac{\bar{r}_k}{1 + \bar{r}_k}\sigma + \frac{\pi/2}{k + 1}\right]$$

$$= A\left(\frac{1}{\alpha} \ln R^n\right)^{(k+1)/k}$$

$$= A\left[\frac{n}{\dfrac{2\pi\mu}{5}\left(\dfrac{n}{n - 1}\right)} \ln \frac{1 + \sin \dfrac{\mu\pi}{40}(\dfrac{n}{n - 1})}{\mu \cdot \sin \dfrac{\mu\pi}{40}(\dfrac{n}{n - 1})}\right]^{(k+1)/k}$$

$$\leq \ M \left[\frac{n(|\ln\mu| + N)}{\mu}\right]^{(k+1)/k},$$

where M and N are constants independent of k, n and μ. This proves the theorem. ¶

Corollary 5.10 *Given $n > 1$ and $0 < \mu < 1$, there exist R, h and $M > 0$, $N > 0$ with the following properties: if $(z_0, f) \in S_R^1 \times P_n(1)$ then there is an l,*

$$l \leq eMn\left[\frac{|\ln\mu| + N}{\mu}\right]^{1+1/\lceil\ln n\rceil},$$

such that $z_l = E_{\lceil\ln n\rceil(h,f)}^l(z_0)$ is an approximate zero of f with probability at least $1 - \mu$.

Proof. Take $k = \lceil\ln n\rceil$. Then Theorem 5.9 gives

$$
\begin{aligned}
l \ &\leq \ M\left[\frac{n(|\ln\mu| + N)}{\mu}\right]^{1+1/\lceil\ln n\rceil} \\
&= \ Mn^{1+1/\lceil\ln n\rceil}\left[\frac{|\ln\mu| + N}{\mu}\right]^{1+1/\lceil\ln n\rceil} \\
&= \ e^{\ln n/\lceil\ln n\rceil}Mn\left[\frac{|\ln\mu| + N}{\mu}\right]^{1+1/\lceil\ln n\rceil} \\
&\leq \ eMn\left[\frac{|\ln\mu| + N}{\mu}\right]^{1+1/\lceil\ln n\rceil}.\ \text{¶}
\end{aligned}
$$

§6. Incremental Algorithms of Efficiency k

With the arguments similar to the proof of Theorem 4.8, now we extend Theorems 4.8 and 4.9 to any incremental algorithm $I_{h,f}$ of efficiency k.

Lemma 6.1 *Suppose that $I_{h,f}$ is an incremental algorithm of efficiency k, i.e., there are $\delta > 0$, $K > 0$, $c_1 > 0$ and $c_2, \cdots, c_k$ independent of h, f and z such that*

$$\frac{f(z')}{f(z)} = \frac{f(I_{h,f}(z))}{f(z)} = 1 - \sum_{j=1}^{k} c_j h^j + S_{k+1}(h),$$

where $|S_{k+1}| \leq Kh^{k+1}\max\{1, 1/h_1^k\}$ and $0 < h < \delta \cdot \min\{1, h_1\}$. Then there exists a constant a with $0 < a \leq 1$ depending only on $I_{h,f}$ such that the following is true: If $0 < h < a \cdot \min\{1, h_1\}$, then

$$\left| \frac{f(z')}{f(z)} \right| < 1 - \frac{c_1 h}{2}$$

and

$$\left| \arg \frac{f(z')}{f(z)} \right| < 2 K h^{k+1} \max\left\{ 1, \frac{1}{h_1^k} \right\}.$$

Proof. Set

$$a = \min\left\{ 1, \delta, \frac{1}{3c_1}, \frac{c_1}{4 \sum_{j=2}^k |c_j|}, \left(\frac{c_1}{4K} \right)^{1/k} \right\},$$

Understanding $c_1/4 \sum_{j=2}^k |c_j| = +\infty$ if $\sum_{j=2}^k |c_j| = 0$, we have

$$0 < h < a \cdot \min\{1, h_1\} \le \delta \cdot \min\{1, h_1\},$$

$$\frac{f(z')}{f(z)} = 1 - c_1 h - \sum_{j=2}^k c_j h^j + S_{k+1}(h) = 1 - c_1 h + \alpha,$$

$$\left| \sum_{j=2}^k c_j h^j \right| \le h^2 \sum_{j=2}^k |c_j| < ha \sum_{j=2}^k |c_j| \le \frac{c_1 h}{4},$$

$$|S_{k+1}(h)| \le hK \max\left\{ h^k, \left(\frac{h}{h_1} \right)^k \right\} \le hKa^k < \frac{c_1 h}{4},$$

$$|\alpha| \le \left| \sum_{j=2}^k c_j h^j \right| + |S_{k+1}(h)| < \frac{hc_1}{4} + \frac{hc_1}{4} = \frac{hc_1}{2},$$

$$\left| \operatorname{Im} \frac{f(z')}{f(z)} \right| = |\operatorname{Im} \alpha| = |\operatorname{Im} S_{k+1}(h)| \le |S_{k+1}(h)|,$$

$$\left| \frac{f(z')}{f(z)} \right| \le (1 - c_1 h) + |\alpha| < (1 - c_1 h) + \frac{c_1 h}{2} = 1 - \frac{c_1 h}{2}.$$

Moreover, since

$$\frac{3}{2} c_1 h < \frac{3}{2} c_1 a \le \frac{1}{2}$$

and

$$0 < 1 - \frac{3}{2} c_1 h < 1 - c_1 h - |\alpha| < \operatorname{Re}\left(\frac{f(z')}{f(z)} \right),$$

we obtain (cf. Fig. 5.12)

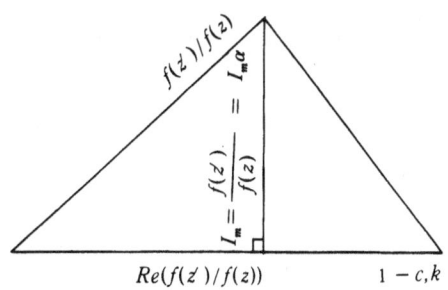

Figure 5.12

$$\left| \arg \frac{f(z')}{f(z)} \right| = \arctan \frac{\left| \mathrm{Im} \dfrac{f(z')}{f(z)} \right|}{\left| \mathrm{Re} \dfrac{f(z')}{f(z)} \right|}$$

$$\leq \frac{\left| \mathrm{Im} \dfrac{f(z')}{f(z)} \right|}{\left| \mathrm{Re} \dfrac{f(z')}{f(z)} \right|}$$

$$\leq \frac{S_{k+1}(h)}{\left| \mathrm{Re} \dfrac{f(z')}{f(z)} \right|}$$

$$\leq \frac{K h^{k+1} \max\left\{ 1, \dfrac{1}{h_1^k} \right\}}{1 - 3c_1 h/2}$$

$$\leq \frac{K h^{k+1} \max\left\{ 1, \dfrac{1}{h_1^k} \right\}}{1 - 1/2}$$

$$= 2 K h^{k+1} \max\left\{ 1, \frac{1}{h_1^k} \right\}. \P$$

Lemma 6.2 *If $0 < \theta \leq \pi/2$ then $1 < \theta/\sin\theta \leq \pi/2$.*

Proof. Let $\varphi(\theta) = \theta/\sin\theta$ for $0 < \theta \leq \pi/2$. Then

$$\lim_{\theta \to 0^+} \varphi(\theta) = 1, \quad \varphi\left(\frac{\pi}{2}\right) = \frac{\pi}{2}.$$

Moreover,

$$\varphi'(\theta) = \left(\frac{\theta}{\sin\theta}\right)' = \frac{\sin\theta - \theta\cos\theta}{\sin^2\theta} = \frac{1}{\sin\theta}\left(1 - \frac{\theta}{\tan\theta}\right) > 0,$$

that is, $\varphi(\theta)$ is strictly increasing. Thus, $1 < \theta/\sin\theta \leq \pi/2$. ¶

Definition 6.3 Suppose that $f(z_0) \neq 0$. Let

$$\Lambda_{f,z_0} = \min_{\substack{\theta, f'(\theta)=0 \\ |f(\theta)| \leq 2|f(z_0)|}} \left\{\frac{\pi}{2}, \left|\arg\frac{f(\theta)}{f(z_0)}\right|\right\}.$$

It is clear that $\Lambda_{f,z_0} \leq \theta_{f,z_0}$, and $\Lambda_{f,z_0} > 0$ implies $f'(z_0) \neq 0$.

Lemma 6.4 Let $\Lambda_{f,z_0} > 0, a$ and $I_{h,f}$ be as in Lemma 6.1, and $h^* = \frac{a}{2}\sin\Lambda_{f,z_0}$. Take $0 < h < h^*$ and $z_l = I_{h,f}^l(z_0)$. Then for any l satisfying

$$0 \leq l - 1 \leq \frac{\Lambda_{f,z_0}}{2} \cdot \frac{1}{2K} \cdot \frac{h^{*k}}{h^{k+1}} \quad \text{and} \quad f(z_{l-1}) \neq 0,$$

we have

$$\Lambda_{f,z_{l-1}} \geq \frac{\Lambda_{f,z_0}}{2} \quad \text{and} \quad \left|\frac{f(z_l)}{f(z_0)}\right| < \left(1 - \frac{c_1 h}{2}\right)^l.$$

Proof. We prove the lemma by induction. When $l = 1$, $\Lambda_{f,z_0} > \Lambda_{f,z_0}/2$. By Lemmas 4.4 and 4.5,

$$\frac{1}{2}\sin\Lambda_{f,z_0} < \sin\frac{1}{2}\Lambda_{f,z_0} < \sin\Lambda_{f,z_0} \leq \sin\theta_{f,z_0} \leq h_1(f, z_0).$$

Thus $h^* = \frac{a}{2}\sin\Lambda_{f,z_0} < a \cdot \min\{1, h_1(f, z_0)\}$. By Lemma 6.1,

$$\left|\frac{f(z_1)}{f(z_0)}\right| < 1 - \frac{c_1 h}{2}.$$

Suppose that the lemma is true when $l = s - 1$, that is,

$$\Lambda_{f,z_{s-2}} \geq \frac{\Lambda_{f,z_0}}{2}$$

and

$$\left|\frac{f(z_{s-1})}{f(z_0)}\right| < \left(1 - \frac{c_1 h}{2}\right)^{s-1}.$$

Next, we will prove that the lemma is also true for $l = s$ as long as

$$0 \leq s - 1 \leq \frac{\Lambda_{f,z_0}}{2} \cdot \frac{1}{2K} \cdot \frac{h^{*k}}{h^{k+1}},$$

that is,

$$\Lambda_{f,z_0} - (s-1)\frac{2Kh^{k+1}}{h^{*k}} \geq \frac{\Lambda_{f,z_0}}{2}.$$

Indeed, with the arguments similar to the case $l = 1$, we get

$$\frac{1}{2}\sin \Lambda_{f,z_0} < \sin \frac{1}{2}\Lambda_{f,z_0} < \sin \Lambda_{f,z_{s-2}} \leq \sin \theta_{f,z_{s-2}} \leq h_1(f, z_{s-2})$$

and

$$h^* = \frac{a}{2}\sin \Lambda_{f,z_0} < a \cdot \min\{1, h_1(f, z_{s-1})\}.$$

Then by Lemma 6.1,

$$\left| \arg \frac{f(z_{s-1})}{f(z_{s-2})} \right| < 2Kh^{k+1} \max\left\{1, \frac{1}{h^{*k}}\right\} = 2K\frac{h^{k+1}}{h^{*k}}.$$

Thus,

$$\begin{aligned}
\Lambda_{f,z_{s-1}} &= \min_{\substack{\theta, f'(\theta)=0 \\ |f(\theta)|\leq 2|f(z_0)|}} \left\{\frac{\pi}{2}, \left|\arg \frac{f(\theta)}{f(z_{s-1})}\right|\right\} \\
&= \min_{\substack{\theta, f'(\theta)=0 \\ |f(\theta)|\leq 2|f(z_0)|}} \left\{\frac{\pi}{2}, \left|\arg \frac{f(\theta)}{f(z_0)} \cdot \frac{f(z_0)}{f(z_1)} \cdots \frac{f(z_{s-2})}{f(z_{s-1})}\right|\right\} \\
&\geq \min\left\{\frac{\pi}{2}, \Lambda_{f,z_0} - (s-1)\frac{2Kh^{k+1}}{h^{*k}}\right\} \geq \frac{\Lambda_{f,z_0}}{2}
\end{aligned}$$

and

$$\begin{aligned}
\left|\frac{f(z_s)}{f(z_0)}\right| &= \left|\frac{f(z_s)}{f(z_{s-1})}\right| \cdot \left|\frac{f(z_{s-1})}{f(z_0)}\right| \\
&< \left(1 - \frac{c_1 h}{2}\right)\left(1 - \frac{c_1 h}{2}\right)^{s-1} \\
&= \left(1 - \frac{c_1 h}{2}\right)^s. \blacksquare
\end{aligned}$$

Theorem 6.5 *Suppose that $I_{h,f}$ is an incremental algorithm of efficiency k. If $\Lambda_{f,z_0} > 0$ and $|f(z_0)| > L > 0$ then let $z_l = I_{h,f}^l(z_0)$, we have*

$|f(z_l)| < L$, *for h, l satisfying*

$$0 < h = \min \left\{ \frac{a}{2} \sin \Lambda_{f,z_0}, \left(\frac{c_1}{8K}\right)^{1/k} \frac{a}{\pi} \left(\frac{\Lambda_{f,z_0}^{k+1}}{\ln \frac{|f(z_0)|}{L}}\right)^{1/k} \right\}$$

and

$$l = \left\lceil \frac{\ln \frac{|f(z_0)|}{L}}{c_1 h/2} \right\rceil$$

$$\leq \max \left\{ \frac{4}{c_1 a \sin \Lambda_{f,z_0}} \ln \frac{|f(z_0)|}{L}, \left(\frac{8K}{c_1}\right)^{1/k} \frac{2\pi}{a c_1} \left(\frac{\ln \frac{|f(z_0)|}{L}}{\Lambda_{f,z_0}}\right)^{(k+1)/k} \right\}.$$

Proof. It is clear that $h^* = \frac{a}{2} \sin \Lambda_{f,z_0} \geq \frac{a}{2} \frac{2}{\pi} \Lambda_{f,z_0} = \frac{a}{\pi} \Lambda_{f,z_0}$. Let $0 < h < h^*$. By Lemma 6.4, it follows that for all l satisfying

$$l - 1 \leq \frac{1}{4K} \left(\frac{a}{\pi}\right)^k \frac{\Lambda_{f,z_0}^{k+1}}{h^{k+1}} \leq \frac{\Lambda_{f,z_0}}{2} \cdot \frac{1}{2K} \frac{h^{*k}}{h^{k+1}},$$

we have

$$\left|\frac{f(z_l)}{f(z_0)}\right| < \left(1 - \frac{c_1 h}{2}\right)^l.$$

Moreover, it is easy to see that

$$0 < h \leq \left(\frac{c_1}{8K}\right)^{1/k} \frac{a}{\pi} \left(\frac{\Lambda_{f,z_0}^{k+1}}{\ln \frac{|f(z_0)|}{L}}\right)^{1/k}$$

is equivalent to

$$0 < h^k \leq \frac{(c_1/8K)\,(a/\pi)^k\, \Lambda_{f,z_0}^{k+1}}{\ln \frac{|f(z_0)|}{L}},$$

and is also equivalent to

$$\frac{1}{4K} \left(\frac{a}{\pi}\right)^k \frac{\Lambda_{f,z_0}^{k+1}}{h^{k+1}} \geq \frac{\ln \frac{|f(z_0)|}{L}}{c_1 h/2}.$$

Thus, if

$$h = \min\left\{h^*, \left(\frac{c_1}{8K}\right)^{1/k} \frac{a}{\pi} \left(\frac{\Lambda_{f,z_0}^{k+1}}{\ln\frac{|f(z_0)|}{L}}\right)^{1/k}\right\}$$

$$= \min\left\{\frac{a}{2}\sin\Lambda_{f,z_0}, \left(\frac{c_1}{8K}\right)^{1/k}\frac{a}{\pi}\left(\frac{\Lambda_{f,z_0}^{k+1}}{\ln\frac{|f(z_0)|}{L}}\right)^{1/k}\right\},$$

then there exists an l such that

$$l - 1 \le \frac{1}{4K}\left(\frac{a}{\pi}\right)^k \frac{\Lambda_{f,z_0}^{k+1}}{h^{k+1}},$$

and

$$l \ge \frac{\ln\frac{|f(z_0)|}{L}}{c_1 h/2} \ge \frac{\ln\frac{|f(z_0)|}{L}}{-\ln(1 - c_1 h/2)}.$$

For example, take

$$l = \left\lceil \frac{\ln\frac{|f(z_0)|}{L}}{c_1 h/2} \right\rceil$$

$$\le \max\left\{\frac{4}{c_1 a \sin\Lambda_{f,z_0}}\ln\frac{|f(z_0)|}{L}, \left(\frac{8K}{c_1}\right)^{1/k}\frac{2\pi}{ac_1}\left(\frac{\ln\frac{|f(z_0)|}{L}}{\Lambda_{f,z_0}}\right)^{(k+1)/k}\right\}.$$

Then

$$l\ln(1 - c_1 h/2) \le \ln\frac{L}{|f(z_0)|},$$

that is,

$$\left(1 - \frac{c_1 h}{2}\right)^l \le \frac{L}{|f(z_0)|}.$$

Therefore,

$$|f(z_l)| < (1 - c_1 h/2)^l |f(z_0)| < L. \P$$

Theorems 6.5 and 4.10 provide criteria for finding approximate zeros by any incremental algorithm of efficiency k.

The discussions in this chapter mainly concern the incremental Euler algorithm $E_{k(h,f)}$, the generalized incremental Euler algorithm $G_{k(h,f)}$ and the incremental algorithm $I_{h,f}$ of efficiency k. For further discussions on the incremental algorithms, refer to [Shub & Smale, 1986], in this paper, M. Shub and S, Smale obtain many interesting results on the Euler-Newton iteration scheme and the generalized Euler algorithm with modification (GEM_k), and also refer to [Xu, 1984].

Chapter 6

Homotopy Algorithms

Let $\mathbf{C}^n = \{z = (z_1, \cdots, z_n) : z_i \in \mathbf{C}, j = 1, \cdots, n\}$, be the n-dimensional complex space. By a polynomial mapping $P : \mathbf{C}^n \to \mathbf{C}^n$, we mean that $P(z) = (P_1(z), \cdots, P_n(z))$ and its every component P_j is a polynomial in $z_1, \cdots, z_n$ with total degree q_j, $j = 1, 2, \cdots, n$. The main idea of homotopy algorithms is to find the zeros of $P(z)$, to choose some trivial polynomial mapping $Q(z)$, and to define a homotopy $H(z, t)$ between $P(z)$ and $Q(z)$ with $H(z, 0) = Q(z)$ and $H(z, 1) = P(z)$, then under some conditions, starting from the zeros of $Q(z)$ and following the zero set

$$H^{-1}(0) = \{(z, t) \in \mathbf{C}^n \times [0, 1] : H(z, t) = 0\},$$

one eventually arrives at the zeros of $P(z)$ in $\mathbf{C}^n \times \{1\}$. Thus, some zeros of $P(z)$ are found.

In sections 1 and 2, we introduce the concepts of homotopy and the degree of continuous mappings, and prove some relevant results such as Index Theorem, Homotopy Invariance Theorem of the degree, etc. This two sections provide necessary preliminaries for understanding homotopy algorithms. Then we obtain some properties of the Jacobian matrices of polynomial mappings in Section 3, and give some conditions to guarantee the boundedness of the zeros of polynomial mappings in Section 4.

The main references of this chapter are [Chow, Mallet-Paret & Yorke, 1978] and [Garcia & Zangwill, 1979a; 1979b; 1979c].

§1. Homotopies and Index Theorem

Definition 1.1 Let $D \subset \mathbf{R}^m$ and $F : D \to \mathbf{R}^m$, $G : D \to \mathbf{R}^m$ be continuous. If $H : D \times [0, 1] \to \mathbf{R}^m$ is continuous with $H(x, 0) = G(x)$

and $H(x,1) = F(x)$ for all $x \in D$, then H is called a homotopy between G and F, and t is called the parameter of the homotopy or the homotopy parameter.

A very simple homotopy between the mappings F and G is the linear homotopy

$$H(x,t) = tF(x) + (1-t)G(x), \quad t \in [0,1].$$

Consider the homotopy equation $H(x,t) = 0$. Let

$$H_t^{-1}(0) = \{(x,t) \in D \times \{t\} : H(x,t) = 0\}, \quad \text{where } t \in [0,1] \text{ is fixed}$$

and

$$H^{-1}(0) = \{(x,t) \in D \times [0,1] : H(x,t) = 0\}.$$

Definition 1.2 Let $D \subset \mathbf{R}^m$ be open and H be a continuously differentiable homotopy. Also let H' be the $m \times (m+1)$ Jacobian matrix of H and H'_x be the $m \times m$ Jacobian matrix of H with respect to x. H is called regular if

(1) the rank of H' is m for every $(x,t) \in H^{-1}(0)$ and

(2) the rank of H'_x is m for every $(x,0) \in H_0^{-1}(0)$ and for every $(x,1) \in H_1^{-1}(0)$.

Remark 1.3 Definition 1.2(2) and Implicit Function Theorem imply that, when $(x,0) \in H_0^{-1}(0)$ (or $(x,1) \in H_1^{-1}(0)$), there exists a neighborhood of $(x,0)$ (or of $(x,1)$) in $\mathbf{R}^m \times \{0\}$ (or in $\mathbf{R}^m \times \{1\}$) such that $(x,0)$ (or $(x,1)$) is the unique element of $H_0^{-1}(0)$ (or of $H_1^{-1}(0)$) lying in the neighborhood. Thus, $H_0^{-1}(0)$ (or $H_1^{-1}(0)$) consists of some isolated points.

Furthermore, denote $t = x_{m+1}$ and let $(x^*, x_{m+1}^*) \in H^{-1}(0)$. Then by Definition 1.2(1) and Implicit Function Theorem, there is some x_j such that

$$x_k(x_j), \quad k = 1, \cdots, j-1, j+1, \cdots, m+1$$

are continuously differentiable functions in x_j and

$$(x_1(x_j), \cdots, x_{j-1}(x_j), x_j, x_{j+1}(x_j), \cdots, x_{m+1}(x_j)) \in H^{-1}(0)$$

with

$$x_k^* = x_k(x_j^*) \qquad \text{for } k = 1, \cdots, j-1, j+1, \cdots, m+1$$

in some neighborhood of (x^*, x_{m+1}^*). In other words, there is a continuously differentiable path (curve) in the neighborhood such that it is the unique solution curve of $H(x, x_{m+1}) = 0$ in the neighborhood.

All of these observations shows that $H^{-1}(0)$ consists of some continuously differentiable paths, and every path can be locally parameterized by some x_j. Of course, we can assign a global parameter for every path, such as its arc length.

Theorem 1.4 *Let $D \subset \mathbf{R}^m$ be an open bounded set and $H : \overline{D} \times [0,1] \to \mathbf{R}^m$ be a regular homotopy. Then $H^{-1}(0)$ consists of a finite number of continuously differentiable paths, and every path is either a loop in $\overline{D} \times (0,1)$ or a path with two ends lying in the boundary of $\overline{D} \times [0,1]$.*

Proof.   The first part of the theorem follows directly from Remark 1.3. It is well-known that every connected and differentiable path is diffeomorphic either to a line segment or to a circle (a loop). Thus the second part of the theorem is also obvious.

Notice that Definition 1.2(1) guarantees that the curves in Fig. 6.1 can't occur in $H^{-1}(0)$, and Definition 1.2(2) makes sure that $H^{-1}(0)$ doesn't contain the curves as shown in Fig. 6.2. The possible curves to appear in $H^{-1}(0)$ are as shown in Fig. 6.3. ¶

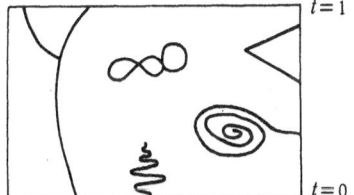
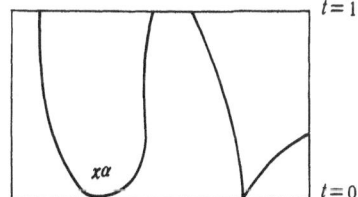

Figure 6.1   Figure 6.2

Definition 1.5   Let $(x^0, x^0_{m+1}) \in H^{-1}(0)$. If $x_i(\theta)$ satisfies

$$(*)\begin{cases} \dfrac{dx_i}{d\theta} = (-1)^{i+1} \det H'_{-i}, \quad i = 1, \cdots, m, m+1 \\ (x(0), x_{m+1}(0)) = (x^0, x^0_{m+1}), \end{cases}$$

then $(x(\theta), x_{m+1}(\theta))$ is called a solution of the basic differential equation $(*)$, where

$$H'_{-j} = \left(\frac{\partial H}{\partial x_1}, \cdots, \frac{\partial H}{\partial x_{j-1}}, \frac{\widehat{\partial H}}{\partial x_j}, \frac{\partial H}{\partial x_{j+1}}, \cdots, \frac{\partial H}{\partial x_{m+1}} \right),$$

here "^" indicates that the corresponding column is deleted.

Theorem 1.6 *Let $D \subset \mathbf{R}^m$ be an open bounded set and $H : \overline{D} \times [0,1] \to \mathbf{R}^m$ be a continuously differentiable homotopy and satisfy Definition 1.2(1). Then every solution of the basic differential equation (∗) is exactly a path in $H^{-1}(0)$.*

Proof. By Definition 1.2(1), there is some j such that H'_{-j} is invertible. Without loss of generality, assume $j = 1$. Since $(x(\theta), x_{m+1}(\theta))$ is a solution of the basic differential equation, then for $i = 2, \cdots, m+1$,

$$\frac{dx_i}{d\theta} = (-1)^{i+1} \det H'_{-i}$$

$$= \frac{1}{\det H'_{-1}} \det \left(\frac{\partial H}{\partial x_2}, \cdots, \frac{\partial H}{\partial x_{i-1}}, -\frac{\partial H}{\partial x_1} \det H'_{-1}, \frac{\partial H}{\partial x_{i+1}}, \cdots, \frac{\partial H}{\partial x_{m+1}} \right).$$

By Cramer rule, the above linear system of equations is equivalent to

$$H'_{-1} \begin{pmatrix} \dfrac{dx_2}{d\theta} \\ \vdots \\ \dfrac{dx_{m+1}}{d\theta} \end{pmatrix} = -\frac{\partial H}{\partial x_1} \det H'_{-1}.$$

Since $\dfrac{dx_1}{d\theta} = \det H'_{-1}$, the above system satisfies

$$\sum_{i=2}^{m+1} \frac{\partial H}{\partial x_i} \frac{dx_i}{d\theta} = -\frac{\partial H}{\partial x_1} \frac{dx_1}{d\theta}.$$

Thus

$$\begin{cases} \dfrac{dH(x(\theta), x_{m+1}(\theta))}{d\theta} = \displaystyle\sum_{i=1}^{m+1} \frac{\partial H}{\partial x_i} \frac{dx_i}{d\theta} = 0 \\ H(x(0), x_{m+1}(0)) = H(x^0, x^0_{m+1}) = 0. \end{cases}$$

This gives $H(x(\theta), x_{m+1}(\theta)) = 0$, that is, $(x(\theta), x_{m+1}(\theta))$ is a path in $H^{-1}(0)$. ¶

Remark 1.7 It is easy to see by a parameter transformation that a path in $H^{-1}(0)$ may not be a solution of the above basic differential equations.

From Theorem 1.6 we can prove

Theorem 1.8 (Index Theorem) *Let $D \subset \mathbf{R}^m$ be an open bounded set and $H : \overline{D} \times [0,1] \to \mathbf{R}^m$ be a regular homotopy. Consider a path of $H^{-1}(0)$ with both ends in $H_0^{-1}(0) \bigcup H_1^{-1}(0)$.*

If the two ends $(x^{c_1}, 0)$ and $(x^{c_2}, 0)$ (or $(x^{c_1}, 1)$ and $(x^{c_2}, 1)$) of the path lie in $H_0^{-1}(0)$ (or $H_1^{-1}(0)$) then

$$\operatorname{sgn} \det H'_{-t}(x^{c_1}, 0) = -\operatorname{sgn} \det H'_{-t}(x^{c_2}, 0)$$

or

$$\operatorname{sgn} \det H'_{-t}(x^{c_1}, 1) = -\operatorname{sgn} \det H'_{-t}(x^{c_2}, 1).$$

If one end $(x^{d_1}, 0)$ of the path lies in $H_0^{-1}(0)$ and the other $(x^{d_2}, 1)$ in $H_1^{-1}(0)$ then

$$\operatorname{sgn} \det H'_{-t}(x^{d_1}, 0) = \operatorname{sgn} \det H'_{-t}(x^{d_2}, 1),$$

where $H'_{-t} = H'_{m+1}$.

Proof. The basic differential equation $(*)$ gives

$$\frac{dt}{d\theta} = (-1)^{m+2} \det H'_{-t}.$$

On the other hand, Definition 1.2(2) implies that $\det H'_{-t} \neq 0$ at the ends. So the corresponding $\dfrac{dt}{d\theta} \neq 0$.

In Fig. 6.3, when $(x^{c_1}, 0)$ moves to $(x^{c_2}, 0)$ along the path C, if the global parameter θ of the path increases then

$$\left.\frac{dt}{d\theta}\right|_{(x^{c_1}, 0)} > 0 \quad \text{and} \quad \left.\frac{dt}{d\theta}\right|_{(x^{c_2}, 0)} < 0.$$

If θ decreases then

$$\left.\frac{dt}{d\theta}\right|_{(x^{c_1}, 0)} < 0 \quad \text{and} \quad \left.\frac{dt}{d\theta}\right|_{(x^{c_2}, 0)} > 0.$$

Thereby,

$$\operatorname{sgn} \det H'_{-t}(x^{c_1}, 0) = -\operatorname{sgn} \det H'_{-t}(x^{c_2}, 0).$$

Consider the path D. When $(x^{d_1}, 0)$ moves to $(x^{d_2}, 1)$ along the path D. Similarly, we have either

$$\left.\frac{dt}{d\theta}\right|_{(x^{d_1}, 0)} > 0 \quad \text{and} \quad \left.\frac{dt}{d\theta}\right|_{(x^{d_2}, 1)} > 0$$

if the parameter θ increases or

$$\left.\frac{dt}{d\theta}\right|_{(x^{d_1}, 0)} < 0 \quad \text{and} \quad \left.\frac{dt}{d\theta}\right|_{(x^{d_2}, 1)} < 0$$

if θ decreases. Thus

$$\text{sgn det } H'_{-t}(x^{d_1}, 0) = \text{sgn det } H'_{-t}(x^{d_2}, 1). \P$$

Remark 1.9   Consider the path A in Fig. 6.3. By Theorem 1.8,

$$\text{sgn det } H'_{-t}(x^a, 0) = -\text{sgn det } H'_{-t}(x^a, 0).$$

This implies that $\det H'_{-t}(x^a, 0) = 0$, and thus contradicts to (2) of Definition 1.2. Therefore, the paths A and G can't appear in $H^{-1}(0)$.

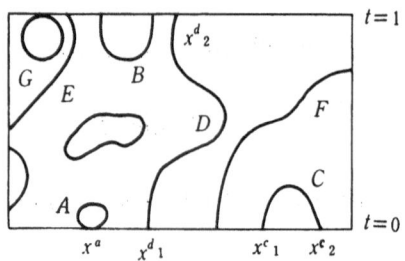

Figure 6.3

Corollary 1.10   *Let $H(x, t) = tF(x) + (1 - t)G(x)$ be a regular homotopy.*

(1) *If a path starts at $(x^a, 1) \in H_1^{-1}(0)$ and ends at $(x^b, 1) \in H_1^{-1}(0)$ then*

$$\text{sgn det } F'(x^a) = -\text{sgn det } F'(x^b).$$

Similar equality is true for the function G.

(2) *If a path starts at $(x^a, 0) \in H_0^{-1}(0)$ and ends at $(x^b, 1) \in H_1^{-1}(0)$ then*

$$\text{sgn det } G'(x^a) = \text{sgn det } F'(x^b).$$

Proof.   (1) It is clear that

$$H'_{-t}(x, 1) = F'(x) \text{ and } H'_{-t}(x, 0) = G'(x).$$

By Theorem 1.8,

$$
\begin{aligned}
\text{sgn det } F'(x^a) &= \text{sgn det } H'_{-t}(x^a, 1) \\
&= -\text{sgn det } H'_{-t}(x^b, 1) \\
&= -\text{sgn det } F'(x^b).
\end{aligned}
$$

(2) Similarly, we obtain

$$
\begin{aligned}
\operatorname{sgn} \det G'(x^a) &= \operatorname{sgn} \det H'_{-t}(x^a, 0) \\
&= \operatorname{sgn} \det H'_{-t}(x^b, 1) \\
&= \operatorname{sgn} \det F'(x^b). \P
\end{aligned}
$$

§2. Degree and its Invariance

In this section, D is always an open bounded set of $\mathbf{R}^m$ unless specified otherwise.

Definition 2.1 Let $F : \overline{D} \to \mathbf{R}^m$ be continuously differentiable. For some $c \in \mathbf{R}^m$, denote

$$
F^{-1}(c) = \{x \in D : F(x) = c\} = \{x \in D : F(x) - c = 0\}.
$$

Call c a regular value of F if the Jacobian matrix F' of F is nonsingular at every point of $F^{-1}(c)$.

Definition 2.2 Let $F : \overline{D} \to \mathbf{R}^m$ be continuously differentiable. If c is a regular value of F then call

$$
\deg(F, \overline{D}, c) = \sum_{x \in F^{-1}(c)} \operatorname{sgn} \det F'(x)
$$

the degree of F with respect to $\overline{D}$ and c.

Since the regularity of c and the compactness of $\overline{D}$ imply that $F^{-1}(c)$ is a finite set, the degree of F is well defined and is an integer.

In this section we will only discuss the linear homotopy

$$
H(x, t) = tF(x) + (1 - t)G(x).
$$

The results obtained in this section are enough for the discussions followed.

Theorem 2.3 *Let $H : \overline{D} \times [0, 1] \to \mathbf{R}^m$ be a continuously differentiable mapping. If for some $c \in \mathbf{R}^m$, $H - c$ is a regular homotopy and $H^{-1}(c) \cap \partial D \times [0, 1] = \emptyset$ then*

$$
\deg(F, \overline{D}, c) = \deg(G, \overline{D}, c),
$$

where $H^{-1}(c) = \{(x, t) \in \overline{D} \times [0, 1] : H(x, t) = c\}$.

Proof.   Since $H-c$ is a regular homotopy and $H^{-1}(c)\cap\partial D\times[0,1]=\emptyset$, $H^{-1}(c)$ consists of paths as shown in Fig. 6.4, and the paths E and F in Fig. 6.3 can't occur.

Notice that $H(x,t)-c=t(F(x)-c)+(1-t)(G(x)-c)$. Apply Corollary 1.10 to the homotopy $H-c$ to obtain following results: If a path links a zero of $F(x)-c$ and a zero of $G(x)-c$ then the corresponding determinants have same sign. If a path links two zeros of $F(x)-c$ (or $G(x)-c$) then the signs of the determinants are opposite. So they do not contribute to the degree sum. Loops such as L in Fig. 6.4 do not enter the sum, so they are irrelevant. Therefore,

$$
\begin{aligned}
\deg(F,\overline{D},c) &= \sum_{x\in F^{-1}(c)} \operatorname{sgn}\det F'(x) \\
&= \sum_{x\in(F-c)^{-1}(0)} \operatorname{sgn}\det(F(x)-c)' \\
&= \sum_{x\in(G-c)^{-1}(0)} \operatorname{sgn}\det(G(x)-c)' \\
&= \deg(G,\overline{D},c).\P
\end{aligned}
$$

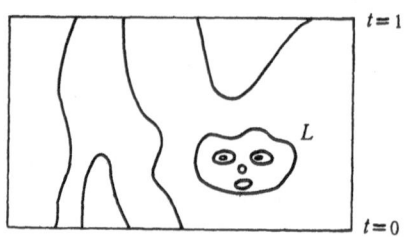

Figure 6.4

To eliminate the hypothesis of the regularity of $H-c$ in Theorem 2.3, we need the following weak continuity of the degree and Sard's Theorem.

Theorem 2.4 *Let $F:\overline{D}\to\mathbf{R}^m$ be a continuously differentiable mapping. Let c be a regular value of F and $F^{-1}(c)\cap\partial D=\emptyset$. Then there exists an $\epsilon>0$ such that when $|e|<\epsilon$ and $c+e$ is a regular value of F,*

$$\deg(F,\overline{D},c+e)=\deg(F,\overline{D},c).$$

Furthermore, $F^{-1}(c+e)\cap\partial D=\emptyset$ for ϵ small enough.

Proof. Notice that $F^{-1}(c)$ is a finite set since $\overline{D}$ is compact and c is a regular value of F. Due to $F^{-1}(c) \cap D = \emptyset$, there is an open neighborhood $N(x^*)$ for every $x^* \in F^{-1}(c)$ such that the restriction of F to $N(x^*)$ is a one-to-one mapping. Choose $\epsilon > 0$ small enough such that $F^{-1}(c + e) \cap [\overline{D} - \bigcup_{x^* \in F^{-1}(c)} N(x^*)] = \emptyset$ as long as $|e| < \epsilon$. Therefore, there exists a one-to-one correspondence between $x^* \in F^{-1}(c)$ and $x^e \in F^{-1}(c + e) \cap N(x^*)$.

It is easy to see that the regularity of c implies $\operatorname{sgn} \det F'(x^*) \neq 0$. Since F is continuously differentiable, if $|e| < \epsilon$ for ϵ small enough then

$$\operatorname{sgn} \det F'(x^e) = \operatorname{sgn} \det F'(x^*).$$

Finally, since $x^* \in D$, we can choose $\epsilon > 0$ small enough such that $x^e \in D$, i.e., $F^{-1}(c + e) \cap \partial D = \emptyset$.

The above discussions give

$$
\begin{aligned}
\deg(F, \overline{D}, c + e) &= \sum_{x^e \in F^{-1}(c+e)} \operatorname{sgn} \det F'(x^e) \\
&= \sum_{x^* \in F^{-1}(c)} \operatorname{sgn} \det F'(x^*) \\
&= \deg(F, \overline{D}, c). \P
\end{aligned}
$$

Definition 2.5 Let $D \subset \mathbf{R}^p$ be open and $F : \overline{D} \to \mathbf{R}^m$ be continuously differentiable. x is called a critical point of F if $\operatorname{rank} F'(x) < m$. Otherwise, x is called a regular point of F. y is called a critical value of F if there exists a critical point x of F such that $y = F(x)$. Otherwise, y is called a regular value.

It is noteworthy that every point in $\mathbf{R}^m - F(\overline{D})$ is automatically a regular value of F, and Definition 2.1 is a special case of Definition 2.5. For the convenience of discussions, denote

$$C_F = \{y \in \mathbf{R}^m : y \text{ is a critical value of } F\},$$

the set of critical values of F.

Theorem 2.6 (Sard's theorem)   Let $D \subset \mathbf{R}^p$ be open, $F : \overline{D} \to \mathbf{R}^m$ be a C^r mapping and $r > \max\{0, p - m\}$. Then C_F has measure zero in $\mathbf{R}^m$.

A proof of Sard's Theorem can be found in [Sternberg, 1964] or [Abraham & Robbin, 1967]. Here the meaning of a C^r mapping is: F is continuous when $r = 0$; F has all r-th continuous partial derivatives when

$0 < r < \infty$; All of its partial derivatives exist and are continuous when $r = \infty$.

Now, we are ready to strengthen Theorem 2.3 and obtain

Theorem 2.7 *Let the homotopy $H : \overline{D} \times [0,1] \to \mathbf{R}^m$ defined by $H(x,t) = tF(x)+(1-t)G(x)$ be a C^2 mapping. If $H^{-1}(c) \cap \partial D \times [0,1] = \emptyset$ and c is a regular value of F and G, then*

$$\deg(F,\overline{D},c) = \deg(G,\overline{D},c).$$

Proof. By Sard's Theorem, there exists e which can arbitrarily approach 0, the origin, such that $c+e$ is a regular value of H. Furthermore, since H is a C^2 mapping, $H^{-1}(c) \cap \partial D \times [0,1] = \emptyset$ and c is a regular value of both F and G, we can choose e small enough such that

$$H^{-1}(c+e) \cap \partial D \times [0,1] = \emptyset$$

and $c + e$ is also a regular value of both F and G. Then Theorem 2.3 means that

$$\deg(F,\overline{D},c+e) = \deg(G,\overline{D},c+e).$$

Now, by Theorem 2.4, for e small enough, we have

$$\deg(F,\overline{D},c+e) = \deg(F,\overline{D},c)$$

and

$$\deg(G,\overline{D},c+e) = \deg(G,\overline{D},c).$$

Therefore,

$$\deg(F,\overline{D},c) = \deg(G,\overline{D},c).\P$$

Theorem 2.7 points out a way to define the degree of continuous mappings.

Definition 2.8 Let $F : \overline{D} \to \mathbf{R}^m$ be continuous and $F^{-1}(c) \cap \partial D = \emptyset$. Then the degree of F with respect to $\overline{D}$ and c is defined by

$$\deg(F,\overline{D},c) = \lim_{k \to \infty} \deg(F^k,\overline{D},c),$$

where all $F^k : \overline{D} \to \mathbf{R}^m$ are C^2 mappings with c as their regular value, and $\{F^k\}$ uniformly converges to F as $k \to +\infty$, i.e., for every $\epsilon > 0$ given, there is a positive integer K such that $\sup_{x \in \overline{D}} |F(x) - F^k(x)| < \epsilon$ for all $k > K$.

It is well-known that every continuous mapping on a compact set can be arbitrarily approximated with C^2 mappings. In fact, by Weierstrass Theorem [Ortega & Rheinboldt, 1970], the mapping can be uniformly approximated with polynomial mappings. With Sard's Theorem, we know that there exists a sequence of C^2 mappings with c as their regular value which uniformly converges to the continuous mapping.

Lemma 2.9 will tell us that the limit in Definition 2.8 does exist and is also finite, and its value is independent of the sequence $\{F^k\}$ chosen. Therefore, the degree $\deg(F, \overline{D}, c)$ is well defined and is an integer.

Besides, it is easy to see that Definition 2.2 coincides with Definition 2.8 by setting $F^k = F$.

Lemma 2.9   *The limit in Definition 2.8 exists and is finite, and its value is independent of the choice of $\{F^k\}$.*

Proof.   Since D is bounded and $F^{-1}(c) \cap \partial D = \emptyset$, there exists $a > 0$ such that

$$|F(x) - c| > a, \quad \forall x \in \partial D.$$

Because $\{F^k\}$ uniformly converges to F, there is a positive integer K such that for $k > K$,

$$|F^k(x) - F(x)| < \frac{a}{2}, \quad \forall x \in \overline{D}.$$

Let

$$\phi(x, t) = tF^k(x) + (1 - t)F^{k+l}(x).$$

It is obvious that

$$
\begin{aligned}
|\phi(x,t) - F(x)| &\leq t|F^k(x) - F(x)| + (1-t)|F^{k+l}(x) - F(x)| \\
&< t \cdot \frac{a}{2} + (1-t)\frac{a}{2} = \frac{a}{2}.
\end{aligned}
$$

Thus, for any $(x, t) \in \partial D \times [0, 1]$,

$$|\phi(x,t) - c| \geq |F(x) - c| - |\phi(x,t) - F(x)| > a - \frac{a}{2} = \frac{a}{2}.$$

This proves $\phi^{-1}(c) \cap \partial D \times [0, 1] = \emptyset$.

By the assumptions of Definition 2.8, $\phi(x, t)$ is a C^2 mapping and c is a regular value of both F^k and F^{k+l}. By Theorem 2.7,

$$\deg(F^k, \overline{D}, c) = \deg(F^{k+l}, \overline{D}, c).$$

This equation means that $\deg(F^k, \overline{D}, c)$ is a constant for $k > K$. Thus, $\lim_{k \to \infty} \deg(F^k, \overline{D}, c)$ exists and is finite.

Finally, suppose that both $\{F^k\}$ and $\{G^k\}$ uniformly converge to F. We define a sequence $\{\tilde{F}^k\}$ by

$$\tilde{F}^k = \begin{cases} F^l, & \text{if } k = 2l - 1, \\ G^l, & \text{if } k = 2l. \end{cases}$$

then $\{\tilde{F}^k\}$ uniformly converges to F and $\lim_{k \to \infty} \deg(\tilde{F}^k, \overline{D}, c)$ exists and is finite, and thus

$$\lim_{k \to \infty} \deg(F^k, \overline{D}, c) = \lim_{k \to \infty} \deg(\tilde{F}^k, \overline{D}, c)$$
$$= \lim_{k \to \infty} \deg(G^k, \overline{D}, c).$$

This completes the proof. ¶

Lemma 2.10   *Let $F, G, F^k, G^k : \overline{D} \to \mathbf{R}^m$ be continuous mappings and $\{F^k\}$ and $\{G^k\}$ uniformly converge to F and G respectively. Define homotopies $H, H^k : \overline{D} \times [0, 1] \to \mathbf{R}^m$ by*

$$H(x, t) = tF(x) + (1 - t)G(x)$$

and

$$H^k(x, t) = tF^k(x) + (1 - t)G^k(x).$$

Then $\{H^k\}$ uniformly converges to H.

Furthermore, if

$$H^{-1}(c) \bigcap \partial D \times [0, 1] = \emptyset$$

then there exists a positive integer K such that for $k > K$

$$(H^k)^{-1}(c) \bigcap \partial D \times [0, 1] = \emptyset.$$

Proof.   It is easy to see that the uniform convergence of $\{F^k\}$ and $\{G^k\}$ implies that $\{H^k\}$ uniformly converges to H. Since $H^{-1}(c) \bigcap \partial D \times [0, 1] = \emptyset$, there is a positive number a such that

$$|H(x, t) - c| > a, \quad \forall (x, t) \in \partial D \times [0, 1].$$

Then the uniform convergence of $\{H^k\}$ to H implies that there exists a positive integer K such that for $k > K$

$$|H^k(x, t) - c| > a/2, \quad \forall (x, t) \in \partial D \times [0, 1].$$

This gives that $(H^k)^{-1}(c) \bigcap \partial D \times [0, 1] = \emptyset$. ¶

Now, we prove the following Homotopy Invariance Theorem of the degree.

Theorem 2.11 *Let $F : \overline{D} \to \mathbf{R}^m$ and $G : \overline{D} \to \mathbf{R}^m$ be continuous and $c \in \mathbf{R}^m$. If $H(x,t) = tF(x) + (1-t)G(x)$ and $H^{-1}(c) \bigcap \partial D \times [0,1] = \emptyset$ then*

$$\deg(F, \overline{D}, c) = \deg(G, \overline{D}, c).$$

Proof. Let F^k, G^k be C^2 mappings with c as their regular value, and $\{F^k\}$ and $\{G^k\}$ uniformly converge to F and G respectively. Define

$$H^k(x,t) = tF^k(x) + (1-t)G^k(x).$$

Since $H^{-1}(c) \bigcap \partial D \times [0,1] = \emptyset$, by Lemma 2.10,

$$(H^k)^{-1}(c) \bigcap \partial D \times [0,1] = \emptyset, \quad \text{for all } k \text{ large enough.}$$

Thus F^k, G^k and H^k satisfy the hypotheses of Theorem 2.7. Therefore,

$$\deg(F^k, \overline{D}, c) = \deg(G^k, \overline{D}, c).$$

By the Definition 2.8, we obtain that

$$
\begin{aligned}
\deg(F, \overline{D}, c) &= \lim_{k \to \infty} \deg(F^k, \overline{D}, c) \\
&= \lim_{k \to \infty} \deg(G^k, \overline{D}, c) \\
&= \deg(G, \overline{D}, c). \P
\end{aligned}
$$

Combining Weierstrass Theorem, Sard's Theorem and Lemma 2.9, one obtain the following result.

Theorem 2.12 *Let $F^k : \overline{D} \to \mathbf{R}^m$ and $F : \overline{D} \to \mathbf{R}^m$ be continuous with $\{F^k\}$ uniformly converging to F. If $c \in \mathbf{R}^m$ satisfies $F^{-1}(c) \bigcap \partial D = \emptyset$, then for k large enough,*

$$(F^k)^{-1}(c) \bigcap \partial D = \emptyset$$

and

$$\lim_{k \to \infty} \deg(F^k, \overline{D}, c) = \deg(F, \overline{D}, c).$$

Proof. The proof of Lemma 2.9 implies that for k large enough,

$$(F^k)^{-1}(c) \bigcap \partial D = \emptyset.$$

Since F^k is continuous, by Weierstrass Theorem, Sard's theorem and Lemma 2.9, there exists a sequence of C^2 mappings $\{G^k\}$ with c as a regular value of G^k such that $|G^k(x) - F^k(x)| < 1/k$ for all $x \in \overline{D}$ and

$$\deg(F^k, \overline{D}, c) = \deg(G^k, \overline{D}, c).$$

It is clear that $\{G^k\}$ uniformly converges to F. By Theorem 2.8,

$$\lim_{k \to \infty} \deg(F^k, \overline{D}, c) = \lim_{k \to \infty} \deg(G^k, \overline{D}, c)$$
$$= \deg(F, \overline{D}, c).¶$$

Next theorem is a direct consequence of Theorem 2.12.

Theorem 2.13 (Continuity of the Degree)   *Let $F : \overline{D} \to \mathbf{R}^m$ be continuous. If $F^{-1}(y) \bigcap \partial D = \emptyset$ then*

$$\lim_{u \to y} \deg(F, \overline{D}, u) = \deg(F, \overline{D}, y).$$

Proof.   For any sequence $u^k \to y$, by Theorem 2.12,

$$\lim_{k \to \infty} \deg(F, \overline{D}, u^k) = \lim_{k \to \infty} \deg(F - u^k, \overline{D}, 0)$$
$$= \deg(F - y, \overline{D}, 0)$$
$$= \deg(F, \overline{D}, y).$$

This proves that

$$\lim_{u \to y} \deg(F, \overline{D}, u) = \deg(F, \overline{D}, y).¶$$

Theorem 2.14 (Existence of Solutions to Continuous Systems of Equations)   *Let $F : \overline{D} \to \mathbf{R}^m$ be continuous. If*

$$F^{-1}(c) \bigcap \partial D = \emptyset \text{ and } \deg(F, \overline{D}, c) \neq 0$$

then the equation

$$F(x) = c$$

has at least one solution in D.

Proof.   Choose a sequence $\{F^k\}$ of C^2 mappings which uniformly converges to F such that c is a regular value of all F^k. By the proof of Lemma 2.9, we have that for k large enough

$$\deg(F^k, \overline{D}, c) = \deg(F, \overline{D}, c) \neq 0.$$

Thus Definition 2.2 guarantees that there exists an $x^k \in D$ such that $F^k(x^k) = c$ for $k = 1, 2, \cdots$. It is evident that the infinite sequence $\{x^k\}$ contained in the compact set $\overline{D}$ has at least one accumulation point. Denote it by x^∞. Without loss of generality, suppose that $x^k \to x^\infty$ as $k \to \infty$. Then

$$
\begin{aligned}
|F(x^\infty) - c| &= |F^k(x^k) - F(x^\infty)| \\
&\leq |F^k(x^k) - F(x^k)| + |F(x^k) - F(x^\infty)| \\
&\to 0 \quad \text{as } k \to \infty.
\end{aligned}
$$

Thus

$$
F(x^\infty) = c.
$$

Finally, $F^{-1}(c) \cap \partial D = \emptyset$ implies $x^\infty \in D$. ¶

In the rest of this section, we will provide an integrative representation for the degree. It is useful in discussing the homotopy algorithms for the zeros of polynomial mappings.

Lemma 2.15 Let $\Omega \subset \mathbf{R}^m$ be an open hyper-rectangle with each facet paralleling some coordinate hyperplane of $\mathbf{R}^m$, and $\mu = f(y)\,dy_1 \wedge \cdots \wedge dy_m$ be a C^∞ m-form on $\mathbf{R}^m$ satisfying $\int_{\mathbf{R}^m} \mu = 0$ and the support $\mathrm{supp}\mu = \overline{\{x \in \mathbf{R}^m : \mu(x) \neq 0\}} \subset \Omega$. Then there exists a C^∞ $(m-1)$-form ω such that $\mathrm{supp}\omega \subset \Omega$ and $\mu = d\omega$.

Proof. We inductively prove that f can be rewritten as

$$
f(y) = \sum_{j=1}^{m} \frac{\partial g_j}{\partial y_j},
$$

where g_j is a C^∞ mapping and $\mathrm{supp}g_j \subset \Omega$ for $j = 1, \cdots, m$.

When $m = 1$. Let $g_1(y) = \int_{-\infty}^{y} f(t)\,dt$. Then $f(y) = \partial g_1 / \partial y$. Since $\mathrm{supp}f = \mathrm{supp}\mu \subset \Omega$ (a line segment) and $\int_{-\infty}^{\infty} f(t)\,dt = \int_{\mathbf{R}^1} \mu = 0$, $\mathrm{supp}g_1 \subset \Omega$.

Now, suppose that the conclusion is true for dimension m. Then for $m+1$, let $y_{m+1} = t$, $(y, t) = (y_1, \cdots, y_m, t)$ and

$$
u(y) = \int_{-\infty}^{\infty} f(y, t)\,dt.
$$

It is clear that $\mathrm{supp}u(y) \subset \Omega_1$, where Ω_1 is the projection of Ω on the coordinate hyperplanes $\mathbf{R}^m$ of $\mathbf{R}^{m+1}$, and

$$
\int_{\mathbf{R}^m} u(y)\,dy_1 \wedge \cdots \wedge dy_m = \int_{\mathbf{R}^m} \left[\int_{-\infty}^{\infty} f(y, t)dt \right] dy_1 \wedge \cdots \wedge dy_m
$$

$$= \int_{\mathbf{R}^{m+1}} f(y, t) \, dy_1 \wedge \cdots \wedge dy_m \wedge dt$$

$$= \int_{\mathbf{R}^{m+1}} \mu = 0.$$

By the inductive hypothesis,

$$u(y) = \sum_{j=1}^{m} \frac{\partial g_j(y)}{\partial y_j}, \quad \text{supp} g_j \subset \Omega_1.$$

Let $\tau(t)$ be a C^∞ function satisfying $\text{supp}\tau \subset \Omega_2$, where Ω_2 is the projection of Ω on the last axis, and

$$\int_{-\infty}^{\infty} \tau(t) \, dt = 1,$$

then,

$$\int_{-\infty}^{\infty} [f(y, t) - \tau(t)u(y)] \, dt = \int_{-\infty}^{\infty} f(y, t) \, dt - \int_{-\infty}^{\infty} \tau(t) \, dt \cdot u(y)$$

$$= u(y) - u(y) = 0.$$

Let

$$g(y, t) = \int_{-\infty}^{t} [f(y, s) - \tau(s)u(y)] \, ds.$$

Then $g(y, t)$ satisfies that

$$\frac{\partial g}{\partial t} = f(y, t) - \tau(t)u(y)$$

and

$$\text{supp} g \subset \Omega.$$

Thus, we have

$$f(y, y_{m+1}) = \frac{\partial g}{\partial y_m}(y, y_{m+1}) + \sum_{j=1}^{m} \frac{\partial g_j(y)}{\partial y_j} \tau(y_{m+1})$$

$$= \frac{\partial g}{\partial y_{m+1}}(y, y_{m+1}) + \sum_{j=1}^{m} \frac{\partial (g_j(y)\tau(y_{m+1}))}{\partial y_j}.$$

This finishes the induction.

Now, we have

$$f(y) = \sum_{j=1}^{m} \frac{\partial g_j}{\partial y_j}(y), \quad \text{supp} g_j \subset \Omega.$$

Let

$$\omega = \sum_{j=1}^{m} (-1)^{j-1} g_j(y) dy_1 \wedge \cdots \wedge dy_{j-1} \wedge dy_{j+1} \wedge \cdots \wedge dy_m.$$

Then

$$d\omega = \sum_{j=1}^{m} \frac{\partial g_j}{\partial y_j} dy_1 \wedge \cdots \wedge dy_m = f(y) dy_1 \wedge \cdots \wedge dy_m = \mu$$

and

$$\text{supp}\omega \subset \Omega. \P$$

In the next lemma, $\mu \circ F$ denotes the composition of the mappings F and μ.

Lemma 2.16   *Let $\mu_j = f_j(y) dy_1 \wedge \cdots \wedge dy_m$ be a C^∞ m-form on $\mathbf{R}^m$ satisfying $\int_{\mathbf{R}^m} \mu_j = 1$ and $\text{supp}\mu_j \subset \Omega_j$, where $\Omega_j \subset \mathbf{R}^m$ is an open hyper-rectangle with each of its faces paralleling with some coordinate $(m-1)$-hyperplane of $\mathbf{R}^m$, $j = 1, 2$, and $\Omega_1 \cap \Omega_2 \neq \emptyset$. If $D \subset \mathbf{R}^m$ is an open bounded set and $F : \overline{D} \to \mathbf{R}^m$ is a C^1 mapping with $F(\partial D) \cap \Omega_j = \emptyset$ for $j = 1, 2$. Then*

$$\int_D \mu_1 \circ F = \int_D \mu_2 \circ F,$$

that is,

$$\int_D f_1 \circ F(x) \cdot \det F'(x) dx_1 \wedge \cdots \wedge dx_m$$
$$= \int_D f_2 \circ F(x) \cdot \det F'(x) dx_1 \wedge \cdots \wedge dx_m.$$

Proof.   We first discuss the case $\Omega_1 = \Omega_2$. Since

$$\int_{\mathbf{R}^m} (\mu_1 - \mu_2) = \int_{\mathbf{R}^m} \mu_1 - \int_{\mathbf{R}^m} \mu_2 = 1 - 1 = 0,$$

by Lemma 2.15, we obtain that there is a C^∞ $(m-1)$-form ω such that

$$\mu_1 - \mu_2 = d\omega, \quad \text{supp}\omega \subset \Omega_1 = \Omega_2.$$

Then Stokes' Theorem gives

$$\int_D \mu_1 \circ F - \int_D \mu_2 \circ F = \int_D (\mu_1 - \mu_2) \circ F$$

$$= \int_D d\omega \circ F$$

$$= \int_D d(\omega \circ F)$$

$$= \int_{\partial D} \omega \circ F$$

$$= 0.$$

Therefore,

$$\int_D \mu_1 \circ F = \int_D \mu_2 \circ F.$$

In the general case, since $\Omega_1 \cap \Omega_2 \neq \emptyset$, there exists an open hyper-rectangle $\Omega \subset \Omega_1 \cap \Omega_2$ with each of its facets paralleling with some coordinate $(m-1)$-hyperplane of $\mathbf{R}^m$. Choose a C^∞ m-form μ on $\mathbf{R}^m$ such that $\int_{\mathbf{R}^m} \mu = 1$ and $\mathrm{supp}\mu \subset \Omega$. With the result of the case $\Omega_1 = \Omega_2$ we have

$$\int_{\Omega_1} \mu_1 \circ F \;=\; \int_{\Omega_1} \mu \circ F$$

$$= \int_\Omega \mu \circ F$$

$$= \int_{\Omega_2} \mu \circ F$$

$$= \int_{\Omega_2} \mu_2 \circ F. \P$$

Theorem 2.17 (Integrative Representation of the Degree)   *Let* $F :$ $\overline{D} \to \mathbf{R}^m$ *be a* C^1 *mapping,* $c \in \Omega \subset \mathbf{R}^m - F(\partial D)$ *be a regular value of* F, *and* μ *be a* C^∞ *m-form on* $\mathbf{R}^m$ *with* $\mathrm{supp}\mu \subset \Omega$ *and* $\int_{\mathbf{R}^m} \mu = 1$. *Then*

$$\deg(F, \overline{D}, c) = \int_D \mu \circ F,$$

where Ω *is an open hyper-rectangle with each of its facets paralleling with some coordinate* $(m-1)$-*hyperplane of* $\mathbf{R}^m$.

Proof.   Suppose that $F^{-1}(c) = \{x^1, \cdots, x^k\}$. There is an open neighborhood U_j of x^j for $j = 1, 2, \cdots, k$ such that $U_1, \cdots, U_k$ don't intersect each other and the restriction of F on every U_j is a one-to-one mapping. Let

$$\tilde{\Omega} \subset \bigcap_{j=1}^k F(U_j) - F\left(\overline{D} - \bigcup_{j=1}^k U_j\right)$$

be an open hyper-rectangle containing c with each of its facets paralleling with some coordinate $(m-1)$-hyperplane of $\mathbf{R}^m$. Suppose that $\tilde{\mu}$ is a C^∞ m-form on $\mathbf{R}^m$ satisfying $\operatorname{supp}\tilde{\mu} \subset \tilde{\Omega}$ and $\int_{\mathbf{R}^m} \tilde{\mu} = 1$. By Lemma 2.16, we have

$$
\begin{aligned}
\deg(F, \overline{D}, c) &= \sum_{j=1}^{k} \operatorname{sgn} \det F'(x^j) \\
&= \sum_{j=1}^{k} \operatorname{sgn} \det F'(x^j) \int_{\mathbf{R}^m} \tilde{\mu} \\
&= \sum_{j=1}^{k} \int_{U_j} \tilde{\mu} \circ F \\
&= \int_D \tilde{\mu} \circ F = \int_D \mu \circ F. \P
\end{aligned}
$$

§3. Jacobian of Polynomial Mappings

Let $\mathbf{C}^n = \{(z_1, \cdots, z_n) : z_j \in \mathbf{C}, j = 1, \cdots, n\}$, the n-dimensional complex space. Further let $P : \mathbf{C}^n \to \mathbf{C}^n$ be a complex analytic mapping. Each component of P is a complex analytic function in n complex variables. It is clear that polynomial mappings are analytic.

Let $\bar{c}$ denote its conjugate for every complex number c. Let $z_j = x_j + i y_j$, where x_j, y_j are real numbers, $j = 1, \cdots, n$. Denote $x = (x_1, \cdots, x_n)$, $y = (y_1, \cdots, y_n)$ and $u = (x_1, \cdots, x_n, y_1, \cdots, y_n) = (x, y)$. Let $P = (P_1, \cdots, P_n)$ and for $j = 1, \cdots, n$,

$$
P_j(z) = f_j(z) + i\, g_j(z) = f_j(x, y) + i\, g_j(x, y),
$$

where f_j and g_j are real-value functions. Then it is easy to see that

$$
\frac{\partial P_j}{\partial z_l} = \frac{\partial P_j}{\partial x_l} = \frac{\partial f_j}{\partial x_l} + i \frac{\partial g_j}{\partial x_l}
$$

and

$$
\frac{\partial P_j}{\partial z_l} = \frac{\partial P_j}{i\, \partial y_l} = \frac{\partial g_j}{\partial y_l} - i \frac{\partial f_j}{\partial y_l}.
$$

Thus we obtain the Cauchy-Riemann equation

$$
\begin{cases}
\dfrac{\partial f_j}{\partial x_l} = \dfrac{\partial g_j}{\partial y_l}, \\[2mm]
\dfrac{\partial f_j}{\partial y_l} = -\dfrac{\partial g_j}{\partial x_l},
\end{cases}
\qquad j, l = 1, \cdots, n.
$$

Naturally, we can regard P as a real mapping F in $2n$ real variables in the following way:

$$
(z_1, \cdots, z_n) \in \mathbf{C}^n \longleftrightarrow (x_1, \cdots, x_n, y_1, \cdots, y_n) \in \mathbf{R}^{2n},
$$

$$
P = (P_1, \cdots, P_n) \longleftrightarrow F = (f_1, \cdots, f_n, g_1, \cdots, g_n)
$$

and

$$
P : \mathbf{C}^n \to \mathbf{C}^n \longleftrightarrow F : \mathbf{R}^{2n} \to \mathbf{R}^{2n}.
$$

Now, we study the relationship between the 1 matrix of P and that of F. For this purpose, denote the $n \times n$ matrices

$$
A = \left(\frac{\partial f_j}{\partial x_l} \right) = \left(\frac{\partial g_j}{\partial y_l} \right)
$$

and

$$
B = \left(-\frac{\partial f_j}{\partial y_l} \right) = \left(\frac{\partial g_j}{\partial x_l} \right).
$$

Then the Jacobian matrices of P and F can be respectively rewritten as

$$
M = \left(\frac{\partial P_j}{\partial z_l} \right) = \left(\frac{\partial f_j}{\partial x_l} + i \frac{\partial g_j}{\partial x_l} \right) = A + i B
$$

and

$$
N = \begin{pmatrix} \left(\dfrac{\partial f_j}{\partial x_l} \right) & \left(\dfrac{\partial f_j}{\partial y_l} \right) \\[3mm] \left(\dfrac{\partial g_j}{\partial x_l} \right) & \left(\dfrac{\partial g_j}{\partial y_l} \right) \end{pmatrix} = \begin{pmatrix} A & -B \\ B & A \end{pmatrix}.
$$

Theorem 3.1 $\det N = |\det M|^2 \geq 0$.

Proof.   Let I be the $n \times n$ identity matrix. Then

$$
\det \begin{pmatrix} A & -B \\ B & A \end{pmatrix} = \det \begin{pmatrix} I & iI \\ 0 & I \end{pmatrix} \begin{pmatrix} A & -B \\ B & A \end{pmatrix} \begin{pmatrix} I & -iI \\ 0 & I \end{pmatrix}
$$

$$
\begin{aligned}
&= \det \begin{pmatrix} A+iB & 0 \\ B & A-iB \end{pmatrix} \\
&= \det(A+iB)\det(A-iB) \\
&= \det(A+iB)\overline{\det(A+iB)} \\
&= |\det(A+iB)|^2. \P
\end{aligned}
$$

Furthermore, we have

Theorem 3.2 *Let $M = A + iB$ be an $n \times n$ complex matrix and*

$$
N = \begin{pmatrix} A & -B \\ B & A \end{pmatrix}
$$

be its corresponding real matrix. Then

$$
\det N = |\det M|^2 \geq 0
$$

and

$$
\mathrm{rank}N = 2\mathrm{rank}M,
$$

where $\mathrm{rank}M$ is the complex rank of M and $\mathrm{rank}N$ is the real rank of N.

Proof. Let α denote the relationship between M and N, i.e., $N = \alpha(M)$. Then it is clear that

$$
\alpha(M_1 M_2) = \alpha(M_1)\alpha(M_2),
$$

$$
\alpha(I) = \begin{pmatrix} I & 0 \\ 0 & I \end{pmatrix}
$$

and

$$
\alpha(M^{-1}) = \alpha(M)^{-1}.
$$

It is well-known that there is a nonsingular matrix U such that

$$
U^{-1}MU = \begin{pmatrix} J_1 & & \\ & \ddots & \\ & & J_k \end{pmatrix},
$$

where

$$
J_j = \begin{pmatrix} \lambda_j & 1 & & & \\ & \lambda_j & 1 & & \\ & & \ddots & \ddots & \\ & & & \ddots & 1 \\ & & & & \lambda_j \end{pmatrix}
$$

is a Jordan matrix, and its only eigenvalue is λ_j with multiplicity k_j, $j = 1, \cdots, k$.

Let $\lambda_j = a_j + i b_j$ and $L_j = \alpha(J_j)$. Then

$$
L_j = \begin{pmatrix}
a_j & -b_j & 1 & 0 & & & & & \\
b_j & a_j & 0 & 1 & & & & & \\
& & a_j & -b_j & 1 & 0 & & & \\
& & b_j & a_j & 0 & 1 & & & \\
& & & & \ddots & & \ddots & & \\
& & & & & \ddots & & 1 & 0 \\
& & & & & & \ddots & 0 & 1 \\
& & & & & & & a_j & -b_j \\
& & & & & & & b_j & a_j
\end{pmatrix},
$$

$j = 1, \cdots, k$.

Since $\det L_j = (a_j^2 + b_j^2)^{k_j}$,

$$
\begin{aligned}
\det N &= \det \alpha(U)^{-1} N \alpha(U) = \det \alpha(U^{-1} M U) \\
&= \det \alpha \begin{pmatrix} J_1 & & \\ & \ddots & \\ & & J_k \end{pmatrix} = \det \begin{pmatrix} L_1 & & \\ & \ddots & \\ & & L_k \end{pmatrix} \\
&= \prod_{j=1}^{k} (a_j^2 + b_j^2)^{k_j} = \left| \prod_{j=1}^{k} \lambda_j^{k_j} \right|^2 \\
&= |\det M|^2 \geq 0.
\end{aligned}
$$

Finally, take

$$
U = \frac{1}{\sqrt{2}} \begin{pmatrix} I & iI \\ iI & I \end{pmatrix}.
$$

It is clear that

$$
U N U^{-1} = \begin{pmatrix} M & 0 \\ 0 & M \end{pmatrix}.
$$

Notice that $\operatorname{rank}\overline{M} = \operatorname{rank} M$, we have

$$
\operatorname{rank} N = 2 \operatorname{rank} M. \blacksquare
$$

§4. Conditions for Boundedness of Solutions

Let $P : \mathbf{C}^n \to \mathbf{C}^n$ and $Q : \mathbf{C}^n \to \mathbf{C}^n$ be analytic. Define the homotopy $\hat{H} : \mathbf{C}^n \times [0,1] \to \mathbf{C}^n$ by

$$\hat{H}(z,t) = tP(z) + (1-t)Q(z).$$

As in the previous section, let $z_j = x_j + iy_j$, $u = (x_1, \cdots, x_n, y_1, \cdots, y_n)$ and

$$\hat{H}_j(z,t) = f_j(z,t) + ig_j(z,t) = f_j(u,t) + ig_j(u,t).$$

Then for the homotopy $\hat{H}$, the corresponding real homotopy $H : \mathbf{R}^{2n} \times [0,1] \to \mathbf{R}^{2n}$ is

$$H = (f_1, \cdots, f_n, g_1, \cdots, g_n).$$

Lemma 4.1 *If $H : \mathbf{R}^{2n} \times [0,1] \to \mathbf{R}^{2n}$ is a regular C^1 homotopy then for any path $(u(\theta), t(\theta))$ in $H^{-1}(0)$, we have either*

$$\mathrm{sgn}\dot{t} = \mathrm{sgn}\det H'_u(u(\theta), t(\theta)), \quad \text{for all } \theta$$

or

$$\mathrm{sgn}\dot{t} = -\mathrm{sgn}\det H'_u(u(\theta), t(\theta)), \quad \text{for all } \theta.$$

where $\dot{t} = \dfrac{dt}{d\theta}$.

Proof. First, we prove that $\dot{t} = 0$ if and only if $\det H'_u = 0$. Suppose that $\dot{t} = 0$. Then $(\dot{u}, \dot{t}) \neq 0$ implies $\dot{u} \neq 0$. Since

$$H(u(\theta), t(\theta)) = 0$$

and

$$H'_u \dot{u} + H'_t \dot{t} = H'_u \dot{u} = 0,$$

we have $\mathrm{rank} H'_u < m$, and thus $\det H'_u = 0$.

Conversely, if $\det H'_u = 0$ then $\mathrm{rank} H'_u < m$. The regularity of H means $\mathrm{rank} H'_{(u,t)} = m$. Hence $\mathrm{rank} H'_u = m - 1$. Without loss of generality, suppose that the first $m-1$ columns of H'_u is linear independent. Denote by B the square matrix obtained from deleting the m-th column of (H'_u, H'_t). Then B is nonsingular and

$$0 = \det H'_u = \det B \det(B^{-1} H'_u) = \det B \det \begin{pmatrix} I_{m-1} & B^{-1}h_m \\ 0 & \end{pmatrix},$$

where h_m is the last column of H'_u. Then the last component of the vector $B^{-1}h_m$ is zero, i.e., $(B^{-1}h_m)_m = 0$. Since

$$0 = B^{-1}(H'_u\dot{u} + H'_t\dot{t}) = \begin{pmatrix} I_{m-1} & \\ & B^{-1}h_m \\ 0 & \end{pmatrix}\dot{u} + \begin{pmatrix} 0 \\ \vdots \\ 0 \\ 1 \end{pmatrix}\dot{t},$$

$$(B^{-1}h_m)_m\dot{u}_m + \dot{t} = 0,$$

and thus

$$\dot{t} = -(B^{-1}h_m)_m\dot{u}_m = 0.$$

Finally, suppose that $\dot{t} \neq 0$. Then $\det H'_u \neq 0$. Define

$$A(\theta) = \begin{pmatrix} H'_u(u(\theta), t(\theta)) & H'_t(u(\theta), t(\theta)) \\ \dot{u}(\theta)^T & \dot{t}(\theta) \end{pmatrix}$$

and

$$C(\theta) = \begin{pmatrix} H'_u(u(\theta), t(\theta)) & \dot{u}(\theta) \\ 0 & \dot{t}(\theta) \end{pmatrix},$$

where T represents the transpose of matrices.

Since $\operatorname{rank}(H'_u, H'_t) = m$ and the nonzero vector $(\dot{u}, \dot{t})$ is orthogonal to every column of (H'_u, H'_t), so $\operatorname{rank} A(\theta) = m + 1$. This proves that $\det A(\theta) > 0$ for all θ or $\det A(\theta) < 0$ for all θ.

Now, notice that $(\dot{u}, \dot{t}) \neq 0$ and $\det H'_u \neq 0$, and we have

$$\det A(\theta) \cdot \dot{t}(\theta) \cdot \det H'_u(u(\theta), t(\theta)) = \det A(\theta)C(\theta)$$

$$= (\dot{u}(\theta)^T, \dot{t}(\theta))\begin{pmatrix} \dot{u}(\theta) \\ \dot{t}(\theta) \end{pmatrix}(\det H'_u(u(\theta), t(\theta)))^2 > 0.$$

Thereby, we have either $\dot{t}(\theta) \cdot \det H'_u(u(\theta), t(\theta)) > 0$ for all $\dot{t}(\theta) \neq 0$ or $\dot{t}(\theta) \cdot \det H'_u(u(\theta), t(\theta)) < 0$ for all $\dot{t}(\theta) \neq 0$.

The above discussions give the result of this theorem. ¶

Theorem 4.2 Let $\hat{H} : \mathbf{C}^n \times [0, 1] \to \mathbf{C}^n$ be an analytic homotopy and denote by H its corresponding real homotopy. If H is a regular homotopy then

$$\det H'_u(u, t) > 0 \quad \text{for all} \quad (u, t) \in H^{-1}(0)$$

and for any path $(u(\theta), t(\theta))$ *in* $H^{-1}(0)$, *we have either*

$$\dot{t}(\theta) > 0, \quad \text{for all } \theta$$

or

$$\dot{t}(\theta) < 0, \quad \text{for all } \theta.$$

Proof. Since $\hat{H}(z,t)$ is analytic, Theorem 3.1 says

$$\det H'_u(u,t) \geq 0,$$

and the regularity of H gives $\text{rank} H'_{(u,t)} = 2n$. For $(u,t) \in H^{-1}(0)$, if $\det H'_u(u,t) = 0$ then Theorem 3.2 implies $\text{rank} H'_u \leq 2n - 2$. Thus $\text{rank} H'_{(u,t)} \leq 2n - 1$. This contradiction gives that $\det H'_u(u,t) > 0$ for all $(u,t) \in H^{-1}(0)$.

By Lemma 4.1, $\text{sgn}\dot{t} = \text{sgn} \det H'_u(u(\theta), t(\theta)) = 1$ for all θ or $\text{sgn}\dot{t} = \text{sgn} \det H'_u(u(\theta), t(\theta)) = -1$ for all θ. This completes the proof. ¶

Remark 4.3 In the proof of Theorem 4.2, $\det H'_u(u,t) > 0$ implies $\dot{t}(\theta) \neq 0$ for all θ. Thus $\dot{t}(\theta) > 0$ for all θ or $\dot{t}(\theta) < 0$ for all θ. Suppose otherwise that $\dot{t}(\theta)$ changes sign. Then there is some θ_0 such that $\dot{t}(\theta_0) = 0$. This contradicts to $\dot{t}(\theta) \neq 0$ for all θ.

The fact that $\dot{t}(\theta)$ doesn't change sign also implies that t is strictly monotonously increasing in θ. Thus the paths A, B, C, D, E and F as shown in Fig. 6.5 can't occur in $H^{-1}(0)$. The only possible paths in $H^{-1}(0)$ are L, M and N as shown in Fig. 6.1. Notice that Theorem 1.8 and $\det H'_u(u,t) > 0$ also guarantee that the paths B, C and E can't occur in $H^{-1}(0)$.

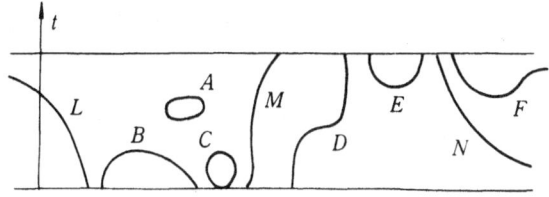

Figure 6.5

Furthermore, the proof of Theorem 4.2 indicates that

H is a regular homotopy

$\Longleftrightarrow \quad \det H'_u(u,t) = |\det \hat{H}'_z(z,t)|^2 > 0$ for all $(z,t) \in H^{-1}(0)$

$\Longleftrightarrow \quad \det H'_z(z,t) \neq 0$ for all $(z,t) \in H^{-1}(0)$.

Now, suppose that $P : \mathbf{C}^n \to \mathbf{C}^n$ is a polynomial mapping. To locate the solutions of the equation $P(z) = 0$, one can start from the known solutions of some trivial system $Q(z) = 0$, and by the homotopy transformation, try to reach the solutions of $P(z) = 0$. That is, in the view of a dynamical geometry, starting from the zeros of $Q(z)$, one follows the paths in $H^{-1}(0)$ and then arrives at the zeros of $P(z)$.

As usual, we can choose $Q_j(z) = z_j^{q_j} - b_j^{q_j}$, where $b_j \neq 0$ and q_j is the degree of P_j for $j = 1, \cdots, n$. It is clear that $Q(z)$ has exactly $q = \prod_{j=1}^n q_j$ zeros and they are

$$(b_1 e^{i\frac{2k_1\pi}{q_1}}, \cdots, b_n e^{i\frac{2k_n\pi}{q_n}}), \quad k_j = 1, \cdots, q_j; \quad j = 1, \cdots, n.$$

Definition 4.4 Let $P(z)$ and $Q(z)$ be polynomial mappings and $\hat{H}(z, t) = tP(z) + (1-t)Q(z)$. P is said to satisfy condition (A) for boundedness of solutions with respect to Q if

(1) $Q^{-1}(0) = \{z | Q(z) = 0\}$ is bounded;

(2) For any sequence $\{z^k\}$ with $|z^k| \to \infty$ there exist some l and a subsequence $\{z^{k_j}\}$ of $\{z^k\}$ satisfying $Q_l(z^{k_j}) \neq 0$ such that

$$\lim_{j\to\infty} \frac{P_l(z^{k_j})}{Q_l(z^{k_j})} \not< 0.$$

The sign $\not< 0$ means that the limit is not a negative real number. If substituting the following (2') for the (2) above, P is said to satisfy condition (B) for boundedness of solutions with respect to Q.

(2') For any sequence $\{z^k\}$ with $|z^k| \to \infty$ there exist some l and a subsequence $\{z^{k_j}\}$ of $\{z^k\}$ satisfying $Q_l(z^{k_j}) \neq 0$ such that

$$\lim_{j\to\infty} \frac{P_l(z^{k_j})}{Q_l(z^{k_j})} \not< \alpha \text{ for some } \alpha > 0.$$

Similarly, the sign $\not< \alpha$ represents that the limit is not a real number less than α.

It is easy to see that the condition (B) implies the condition (A).

Example 4.5 Let P be a polynomial mapping and q_j is the degree of its j-th component, and $Q(z) = (Q_1(z), \cdots, Q_n(z))$, $Q_j(z) = z_j^{q_j+1} - 1$. It is clear that for any sequence $\{z^k\}$ with $|z^k| \to \infty$ there exist some l

and a subsequence $\{z^{k_j}\}$ of $\{z^k\}$ such that

$$\lim_{j\to\infty} \frac{P_l(z^{k_j})}{(z_l^{k_j})^{q_l+1} - 1} = 0,$$

that is, P satisfies the condition (A) for boundedness of solutions with respect to Q.

Example 4.6   Let $P_j(z) = z_j^{q_j} + \hat{P}_j(z)$, $\hat{P}$ be analytic, and $Q_j(z) = z_j^{q_j} - b_j^{q_j}$, $b_j \neq 0$. Suppose that for any sequence $\{z^k\}$ with $|z^k| \to \infty$ there exist some l and a subsequence $\{z^{k_j}\}$ of $\{z^k\}$ such that $|z_l^{k_j}| \to \infty$ and

$$\lim_{j\to\infty} \frac{\hat{P}_l(z^{k_j})}{(z_l^{k_j})^{q_l}} = 0.$$

Then P satisfies the condition (B) for boundedness of solutions with respect to Q. In fact

$$\lim_{j\to\infty} \frac{P_l(z^{k_j})}{Q_l(z^{k_j})} = \lim_{j\to\infty} \frac{(z_l^{k_j})^{q_l} + \hat{P}_l(z^{k_j})}{(z_l^{k_j})^{q_l} - b_l^{q_l}}$$

$$= \lim_{j\to\infty} \left(1 + \frac{\hat{P}_l(z^{k_j})}{(z_l^{k_j})^{q_l}}\right)\left(\frac{(z_l^{k_j})^{q_l}}{(z_l^{k_j})^{q_l} - b_l^{q_l}}\right) = 1.$$

Example 4.7   In Example 4.6, if $\hat{P}_j(z)$ is a polynomial of degree less than q_j, $j = 1, \cdots, n$, then for any sequence $\{z^k\}$ with $|z^k| \to \infty$ there are some l and a subsequence $\{z^{k_j}\}$ satisfying $|z_l^{k_j}| = \max\{|z_1^{k_j}|, \cdots, |z_n^{k_j}|\}$. Thus

$$\lim_{j\to\infty} \frac{\hat{P}_l(z^{k_j})}{(z_l^{k_j})^{q_l}} = \lim_{j\to\infty} \frac{\hat{P}_l(z^{k_j})}{(z_l^{k_j})^{q_l-1}} \cdot \frac{1}{z_l^{k_j}} = 0.$$

This proves that P satisfies the condition (B) with respect to Q.

Next theorem is important for the discussions of homotopy algorithms for zeros of polynomial mappings, especially for the discussions in Chapter 8.

Theorem 4.8 (Boundedness of Solutions)   *Let $P(z)$ and $Q(z)$ be analytic and $\hat{H}(z,t) = tP(z) + (1-t)Q(z)$.*

(1)   *If P satisfies the condition (A) for boundedness of solutions with respect to Q, then for any $0 < \epsilon \leq 1$,*

$$\{(z,t): \hat{H}(z,t) = tP(z) + (1-t)Q(z) = 0, 0 \leq t \leq 1 - \epsilon\}$$

is bounded;

(2)   *If* P *satisfies the condition* (B) *for boundedness of solutions with respect to* Q, *then*

$$\hat{H}^{-1}(0) = \{(z,t) : \hat{H}(z,t) = tP(z) + (1-t)Q(z) = 0\}$$

is bounded.

Proof.   (1) Suppose otherwise that there is some ϵ with $0 < \epsilon \le 1$ such that

$$\{(z,t) : \hat{H}(z,t) = tP(z) + (1-t)Q(z) = 0, 0 \le t \le 1 - \epsilon\}$$

is not bounded, i.e., there is a sequence $\{(z^k, t^k)\}$ with $|z^k| \to \infty$ and $\hat{H}(z^k, t^k) = 0$, $0 \le t^k \le 1 - \epsilon$. Since $Q^{-1}(0)$ is bounded, there is a subsequence $\{(z^{k_j}, t^{k_j})\}$ such that for some l, $Q_l(z^{k_j}) \ne 0$ and

$$t^{k_j} P(z^{k_j}) + (1 - t^{k_j})Q(z^{k_j}) = 0.$$

Thus,

$$0 > \lim_{j \to \infty} \left(1 - \frac{1}{t^{k_j}}\right) = \lim_{j \to \infty} \frac{P_l(z^{k_j})}{Q_l(z^{k_j})} \not< 0.$$

This is a contradiction.

(2) If $\hat{H}^{-1}(0)$ is not bounded, then there is a sequence $\{(z^k, t^k)\}$ with $|z^k| \to \infty$ and $\hat{H}(z^k, t^k) = 0$, $0 \le t^k \le 1$. Since $Q^{-1}(0)$ is bounded, there is a subsequence $\{(z^{k_j}, t^{k_j})\}$ such that for some l, $Q_l(z^{k_j}) \ne 0$ and

$$t^{k_j} P(z^{k_j}) + (1 - t^{k_j})Q(z^{k_j}) = 0.$$

Thus,

$$0 > \lim_{j \to \infty} \left(1 - \frac{1}{t^{k_j}}\right) = \lim_{j \to \infty} \frac{P_l(z^{k_j})}{Q_l(z^{k_j})} \not< \alpha \text{ for some } \alpha.$$

This contradiction implies that $\hat{H}^{-1}(0)$ is bounded. ¶

Remark 4.9 In Theorem 4.2, $\hat{H}$ is a regular homotopy. If in addition, $\hat{H}(z,t) = tP(z) + (1-t)Q(z)$, and P, Q satisfy the condition (A), then the paths A, B, C, D, E, F and N in Fig. 6.5 can't occur. The only possible kinds of paths in $H^{-1}(0)$ are as L and M.

Similarly, if P and Q satisfy the condition (B) then the only possible kind of paths in $\hat{H}^{-1}(0)$ is as M. In this case, the number of zeros of P is equal to that of Q, and is also equal to the number of the paths in $H^{-1}(0)$. In particular, if P and Q are as shown in Example 4.6, then P has exactly $q = \prod_{j=1}^{n} q_j$ zeros.

Chapter 7

Probabilistic Discussion on Zeros of

Polynomial Mappings

Based on Chapter 6, this chapter applies parameterized Sard's Theorem and the degree theory to a probabilistic discussion of the zeros of polynomial mappings.

Theorem 1.13 shows that almost every polynomial mapping has $q = \prod_{j=1}^{n} q_j$ zeros, where q_j is the degree of the i-th component of the mapping. Theorem 1.18 provides a sufficient condition that guarantees the polynomial mappings having exactly q zeros, counting multiplicities.

In Section 2 Theorem 2.4 exhibits some properties of homotopy paths of locating the zeros of polynomial mappings and furnishes a new constructive and topological proof of Bézout Theorem.

Transforming the problem of finding zeros of an analytic function on a bounded domain to that of locating zeros of a special polynomial mapping, Section 3 gives some results on the zeros of analytic functions.

The main references of this chapter are [Garcia & Li, 1980] and [Li, 1983].

§1.   Number of Zeros of Polynomial Mappings

To familiarize with the method of the probabilistic discussions, we first prove a special case of Sard's Theorem, and then discuss probabilistically the number of zeros of polynomials.

Lemma 1.1   *Let $D \subset \mathbf{R}^m$ be open and $F : D \to \mathbf{R}^m$ be a C^1 mapping with $\det F'(x) \equiv 0$. Then $\mathrm{vol}F(D)$, the volume of $F(D)$ in $\mathbf{R}^m$, is zero.*

Proof. It is clear that we need only to prove $\mathrm{vol} F(C_0) = 0$ for every hypercube (of dimension m) C_0 in D

Dividing every edge of C_0 into N equal segments, we obtain N^m small hypercubes in C_0. Let x^0 be a fixed point in some small hypercube. Then for every x in the small hypercube,

$$
\begin{aligned}
F(x) &= F(x^0) + F'(x^0)(x - x^0) + o(|x - x^0|) \\
&= F(x^0) + F'(x^0)(x - x^0) + o\left(\frac{1}{N}\right).
\end{aligned}
$$

Since F is a C^1 mapping and $\det F'(x^0) = 0$, without loss of generality we may assume

$$
F'(x^0) = \begin{pmatrix} & * & \\ 0 \cdots 0 & \end{pmatrix}.
$$

Then

$$
\begin{pmatrix} F_1(x) - F_1(x^0) \\ \vdots \\ F_m(x) - F_m(x^0) \end{pmatrix} = \begin{pmatrix} & * & \\ 0 \cdots 0 & \end{pmatrix} \begin{pmatrix} x_1 - x_1^0 \\ \vdots \\ x_m - x_m^0 \end{pmatrix} + o\left(\frac{1}{N}\right).
$$

Thus there is a constant $K > 0$ such that for every small hypercube, the volume of its image is not larger than

$$
K\left(\frac{1}{N}\right)^{m-1} \cdot o\left(\frac{1}{N}\right) = o\left(\frac{1}{N^m}\right).
$$

This leads to

$$
0 \le \mathrm{vol} F(C_0) \le N^m \cdot o\left(\frac{1}{N^m}\right) \to 0.
$$

Therefore,

$$
\mathrm{vol} F(C_0) = 0.\P
$$

Lemma 1.2 *Let $D \subset \mathbf{R}^{m-1}$ be open and $F : D \to \mathbf{R}^m$ be a C^1 mapping. Then $\mathrm{vol} F(D) = 0$.*

Proof. Define $\tilde{F} : D \times \mathbf{R} \to \mathbf{R}^m$ by

$$
\tilde{F}(x_1, \cdots, x_{m-1}, x_m) := F(x_1, \cdots, x_{m-1}).
$$

It is clear that

$$
\det \tilde{F}'(x_1, \cdots, x_{m-1}, x_m) \equiv 0.
$$

From Lemma 1.1 we obtain

$$\text{vol}F(D) = \text{vol}\tilde{F}(D \times \mathbf{R}) = 0.\P$$

In the following we will continue to discuss the relationship between zeros and coefficients of polynomials.

Lemma 1.3 *Let L be a simple closed curve on the complex plane $\mathbf{C}$ and $P(z)$ be a polynomial whose zeros are not on L. Let D denote the interior of L. If $\xi_1, \cdots, \xi_l$ are the all distinct zeros of P in D and their multiplicities are $k_1, \cdots, k_l$ respectively, then*

$$\frac{1}{2\pi i} \int_L z^k \frac{P'(z)}{P(z)} dz = \sum_{j=1}^{l} k_j \xi_j^k.$$

Proof. In fact, a routine calculation shows that

$$\frac{1}{2\pi i} \int_L z^k \frac{P'(z)}{P(z)} dz$$

$$= \sum_{j=1}^{l} \frac{1}{2\pi i} \int_{L_j} z^k \frac{P'(z)}{P(z)} dz$$

$$= \sum_{j=1}^{l} \frac{1}{2\pi i} \int_{L_j} z^k \frac{k_j(z-\xi_j)^{k_j-1}P_j(z) + (z-\xi_j)^{k_j}P_j'(z)}{(z-\xi_j)^{k_j}P_j(z)} dz$$

$$= \sum_{j=1}^{l} \frac{1}{2\pi i} \int_{L_j} \xi_j^k \frac{k_j}{z-\xi_j} dz$$

$$+ \sum_{j=1}^{l} \frac{1}{2\pi i} \int_{L_j} (z^k - \xi_j^k) \frac{k_j(z-\xi_j)^{k_j-1}P_j(z) + (z-\xi_j)^{k_j}P_j'(z)}{(z-\xi_j)^{k_j}P_j(z)} dz$$

$$= \sum_{j=1}^{l} \frac{1}{2\pi i} \int_0^{2\pi} \xi_j^k \frac{k_j}{re^{i\theta}} rie^{i\theta} d\theta$$

$$= \sum_{j=1}^{l} \frac{1}{2\pi i} \xi_j^k k_j i 2\pi$$

$$= \sum_{j=1}^{l} k_j \xi_j^k,$$

where $L_j \subset D$ is a circle with small radius r surrounding only the zero ξ_j of P (cf. Fig. 7.1). ¶

Remark 1.4   In Lemma 1.3, if $k = 0$ then

$$\frac{1}{2\pi i} \int_L \frac{P'(z)}{P(z)} dz = \sum_{j=1}^{l} k_j$$

gives the number of zeros of P lying in D.

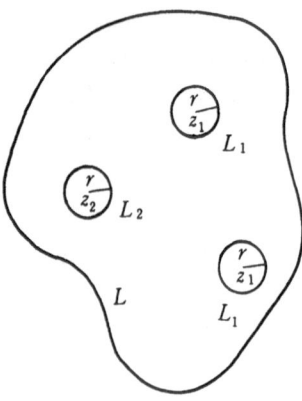

Figure 7.1

We will provide two ways to prove the next theorem.

Theorem 1.5   *Almost every polynomial with degree q has exactly q zeros.*

Proof 1.   Let $P(z) = a_0 z^q + a_1 z^{q-1} + \cdots + a_{q-1} z + a_q$. Then the polynomial is uniquely determined by $(a_0, a_1, \cdots, a_q) \in \mathbf{C}^{q+1}$. This gives a one-to-one relationship between the polynomials of degree at most q and the points in $\mathbf{C}^{q+1}$.

When $a_0 \neq 0$. Then $P(z)$ has multiple zeros if and only if $P(z)$ and $P'(z)$ have common zeros, and also if and only if $R(P, P')$, the resultant of P and P', is zero.

It is clear that $R(\tilde{P}, \tilde{P}') \neq 0$ for the special polynomial $\tilde{P}(z) = z^q - 1$. Thus $R(P, P') \not\equiv 0$. But the measure of the set $\{(a_0, a_1, \cdots, a_n) : R(P, P') = 0\}$ in $\mathbf{C}^{q+1}$ is zero. This says that almost every polynomial with degree q has q distinct zeros. ¶

Proof 2.   Let

$$\begin{aligned}
P(z) &= a_0 z^q + a_1 z^{q-1} + \cdots + a_{q-1} z + a_q \\
&= a_0 (z - \xi_1) \cdots (z - \xi_q) \\
&= a_0 [z^q - (\xi_1 + \cdots + \xi_q) z^{q-1} + \cdots + (-1)^q \xi_1 \cdots \xi_n].
\end{aligned}$$

Then

$$\begin{cases} a_1 = -a_0(\xi_1 + \cdots + \xi_q), \cdots \\ a_q = (-1)^q a_0 \xi_1 \cdots \xi_q, \end{cases}$$

and Lemma 1.2 leads to

$$\text{vol}\{(a_0, a_1, \cdots, a_q) : P \text{ has multiple zeros}\}$$

$$= \text{vol}\{(a_0, a_1, \cdots, a_q) : a_1 = -a_0(\xi_1 + \cdots + \xi_q),$$

$$\cdots, a_q = (-1)^q a_0 \xi_1 \cdots \xi_q, \text{ and there are } j \neq l$$

$$\text{such that } \xi_j = \xi_l\}$$

$$= 0. \P$$

From Lemma 1.3 and Remark 1.4, we have

Theorem 1.6   *The zeros of a polynomial are continuous functions in the coefficients of the polynomial.*¶

This theorem is exactly Lemma 1.4.5.

Theorem 1.7   *Let ξ be a zero of $P(z)$ with multiplicity k and D be an open disk with center ξ and radius r which doesn't contain any other zeros of $P(z)$. Then*

$$\deg(P, \overline{D}, 0) = k.$$

Proof.   By the proof of Theorem 1.5, a polynomial Q has no multiple zeros $\Leftrightarrow$ for any zero ξ of Q, $Q'(\xi) \neq 0 \Leftrightarrow |Q'(\xi)|^2 > 0 \Leftrightarrow 0$ is a regular value of Q. Then by Theorems 1.5 and 1.6, there are polynomials $\{P^l\}$ such that $\{P^l\}$ uniformly converges to P, 0 is a regular value of every P^l and each P^l has exactly k distinct zeros in $\overline{D}$. Thus

$$\deg(P, \overline{D}, 0) = \lim_{l \to \infty} \deg(P^l, \overline{D}, 0) = \lim_{l \to \infty} k = k. \P$$

Now, we turn to discuss the number of the zeros of polynomial mappings in two variables. Next theorem is a special case of Bézout Theorem.

Theorem 1.8   *Let $P_1(z_1, z_2)$ and $P_2(z_1, z_2)$ be polynomials of degrees q_1 and q_2 respectively with $q_1 \geq 1$ and $q_2 \geq 1$. If the number of the common zeros of P_1 and P_2 is greater than $q = q_1 q_2$ then P_1 and P_2 have common factors*

Proof.   Let

$$F_j(z_0, z_1, z_2) = z_0^{q_j} P_j\left(\frac{z_1}{z_0}, \frac{z_2}{z_0}\right), \qquad j = 1, 2.$$

It is clear that F_j is a 1 polynomial of degree q_j, $j = 1, 2$, and the number of the common zeros of F_1 and F_2 is greater than $q = q_1 q_2$. Choose $q_1 q_2 + 1$ distinct common zeros and link every pair of the zeros with a line. Then there is easily a point $z^* = (z_0^*, z_1^*, z_2^*)$ which doesn't lie on the lines and also doesn't lie on the curve $F_j = 0$, $j = 1, 2$. Choose a coordinate system such that $z^* = (0, 0, 0)$. Then

$$F_1(z) = A_0 z_0^{q_1} + A_1 z_0^{q_1 - 1} + \cdots + A_{q_1},$$

$$F_2(z) = B_0 z_0^{q_2} + B_1 z_0^{q_2 - 1} + \cdots + B_{q_2},$$

where A_0 and B_0 are nonzero constants, and A_j, B_j for $j > 0$ are homogeneous polynomials of degree j in z_1, z_2. It is well-known that $R(z_1, z_2)$, the resultant of F_1 and F_2 in z_0, is either 0 or a homogeneous polynomial of degree $q = q_1 q_2$ in z_1, z_2. Moreover, $R(c_1, c_2) = 0$ if and only if there is a c_0 such that $F_1(c) = F_2(c) = 0$, where $c = (c_0, c_1, c_2)$. This means that the last two coordinates c_1, c_2 of every common zero of F_1 and F_2 satisfy $R(c_1, c_2) = 0$. On the other hand, notice that $(1,0,0)$ doesn't lie on any line determined by a pair of the chosen $q_1 q_2 + 1$ distinct common zeros, the ratios $c_1 : c_2$ for all the common zeros (c_0, c_1, c_2) are distinct each other. This means that the number of the zeros of $R(z_1, z_2)$ is at least $q_1 q_2 + 1$, and thus $R(z_1, z_2) \equiv 0$. Therefore, F_1 and F_2 have common factors. (cf. [Walker, 1950]). ¶

Corollary 1.9 *Let $P_1(z_1, z_2)$ and $P_2(z_1, z_2)$ be polynomials of degrees q_1 and q_2 respectively. If P_1 and P_2 have no common factors then the number of the common zeros of P_1 and P_2 is at most $q = q_1 q_2$.* ¶

Now, we start to generalize Theorem 1.5 to the case of general polynomial mappings. We first give

Theorem 1.10 (Parameterized Sard's Theorem) *Let $U \subset \mathbf{R}^p$ and $V \subset \mathbf{R}^s$ be open sets. Also let $F : U \times V \to \mathbf{R}^m$ be a C^r mapping with $r > \max\{0, p - m\}$. If $y \in \mathbf{R}^m$ is a regular value of F then for almost every $v \in V$ (except a subset of V with measure 0), y is a regular value of $F(\cdot, v)$.* ¶

This theorem is a special case of the Transversality Theorem in [Abrabam & Robbins, 1967].

Let $P(z; \omega) : \mathbf{C}^n \times \mathbf{C}^k \to \mathbf{C}^n$ be a mapping such that for every fixed $\omega \in \mathbf{C}^k$, $P(\cdot; \omega)$ is a polynomial mapping whose i-th component P_i is of degree q_i, $i = 1, \cdots, n$, where ω is a vector of the coefficients of P which uniquely defines P and vice versa. We further distinguish the coefficients of P as follows. Let $\omega = (a, b, c, d)$, where $a = (a_{ij}) \in \mathbf{C}^{n^2}$, $b = (b_i) \in \mathbf{C}^n$

and a_{ij} is the coefficient of $z_j^{q_i}$ in P_i, b_i is the constant term of P_i. c is the vector whose components are the coefficients of terms of degree q_i in P_i for all i excepting the a_{ij}'s, and d consists of the rest of the coefficients of P.

Define $H : \mathbf{C}^n \times \mathbf{R} \times \mathbf{C}^k \to \mathbf{C}^n$ by

$$H_i(z, t; \omega) = tP_i(z, \omega) + (1 - t)(z_i^{q_i} - 1), \quad i = 1, \cdots, n.$$

Denote

$$H_\omega^{-1}(0) = \{(z, t) \in \mathbf{C}^n \times [0, 1] : H(z, t; \omega) = 0\}.$$

Lemma 1.11 *For all a, c, d and for almost all b (except a set in $\mathbf{C}^n$ with measure 0), $0 \in \mathbf{C}^n$ is a regular value of both $P(\cdot; \omega)$ and $H(\cdot, \cdot; \omega)$.*

Proof. Since $P_b'(z; a, b, c, d) = I$, Theorem 1.10 means that 0 is a regular value of $P(\cdot; a, b, c, d)$ for all a, c, d and for almost all b.

Similarly, since

$$H_b'(z, t; a, b, c, d) = tI, \quad \text{when } t \neq 0$$

and

$$H_z'(z, t; a, b, c, d) = \begin{pmatrix} q_1 z_1^{q_1-1} & & \\ & \ddots & \\ & & q_n z_n^{q_n-1} \end{pmatrix}, \quad \text{when } t = 0,$$

Theorem 1.10 gives that 0 is a regular value of $H(\cdot, \cdot; a, b, c, d)$ for all a, c, d and for almost all b. ¶

Let P_i^* denote the polynomial consisted of the terms of degree q_i in P_i, $i = 1, \cdots, n$. It is clear that (a, c) uniquely determines P^*. So P^* can be rewritten as $P^*(z; a, c)$. Define

$$H_i^*(z, t; a, c) = tP_i^*(z; a, c) + (1 - t)z_i^{q_i}, \quad i = 1, \cdots, n.$$

Lemma 1.12 *For all c and for almost all $a \in \mathbf{C}^{n^2}$, 0 is a regular value of $H^*(\cdot, \cdot; a, c)$ on its support $\{(z, t) : z \neq 0\}$.*

Proof. When $t = 0$. Then $z = 0$ is the unique zero of $H^*(\cdot, 0; a, c) = 0$.

When $t \neq 0$ and $z \neq 0$. We can choose a j such that $z_j \neq 0$. Then the derivatives of H^* with respect to $a_{\cdot j} = (a_{1j}, \cdots, a_{nj})$ are

$$H_{a_{\cdot j}}^{*'} = \begin{pmatrix} tz_j^{q_1} & & \\ & \ddots & \\ & & tz_j^{q_n} \end{pmatrix}.$$

Thus $H_{a.j}^{*'}$ is nonsingular, and therefore, 0 is a regular value of $H^*(\cdot, \cdot; a, c)$ on $\{(z, t) : z \neq 0\}$ for almost all $a \in \mathbf{C}^{n^2}$. ¶

Now we are ready to prove the main result of this section.

Theorem 1.13 *For almost every fixed ω, $P(z, \omega)$ has exactly $q = \prod_{i=1}^n q_i$ distinct zeros, and $H_\omega^{-1}(0)$ consists of q bounded differentiable paths which don't intersect each other and every path links a zero of the polynomial mapping $H(\cdot, 0; \omega)$ in $\mathbf{C}^n \times \{0\}$ and a zero of $H(\cdot, 1; \omega) = P(\cdot; \omega)$ in $\mathbf{C}^n \times \{1\}$.*

Proof. Lemmas 1.11 and 1.12 show that for almost every ω, i.e., except for ωs in a subset of $\mathbf{C}^k$ of measure zero, 0 is a regular value of both $P(\cdot; \omega)$ and $H(\cdot, \cdot; \omega)$, and 0 is a regular value of $H^*(\cdot, \cdot, \cdot; a, c)$ on $\{(z, t) : z \neq 0\}$.

Let ω be outside the above subset of measure zero. Then Theorem 6.4.2 and Remark 6.4.3 guarantee that $H_\omega^{-1}(0)$ consists of some differentiable paths. Thus it suffices to show that none of the paths diverges to infinity, that is, the paths A, B and C as shown in Fig. 7.2 can't occur in $H_\omega^{-1}(0)$. If this is true, then every path links a trivial zero of $H(\cdot, 0; \omega)$ and a zero of $H(\cdot, 1; \omega) = P(\cdot; \omega) = 0$. Since $H(\cdot, 0; \omega)$ has q distinct zeros, thus $P(\cdot; \omega)$ must have q distinct zeros.

Suppose otherwise that some path $(z(\theta), t(\theta))$ in $H_\omega^{-1}(0)$ diverges to infinity as θ approaches $\bar{\theta}$. For any i, notice that H_i^* is homogeneous in z of degree q_i, we have

$$
\begin{aligned}
H_i^*\left(\frac{z(\theta)}{|z(\theta)|}, t(\theta); a, c\right) &= |z(\theta)|^{-q_i} H_i^*(z(\theta), t(\theta); a, c) \\
&= |z(\theta)|^{-q_i}[H_i^*(z(\theta), t(\theta); a, c) - H_i(z(\theta), t(\theta); \omega)] \\
&= |z(\theta)|^{-q_i}[1 - t(\theta) + t(\theta)(P_i^*(z(\theta); a, c) - P_i(z(\theta); \omega))] \\
&\to 0 \quad \text{as } \theta \to \bar{\theta}.
\end{aligned}
$$

Let (z^0, t^0) be an accumulation point of $\{(z(\theta)/|z(\theta)|, t(\theta))\}$. Then

$$
H^*(z^0, t^0; a, c) = 0, \quad |z^0| = 1, \quad 0 \leq t^0 \leq 1.
$$

Since H^* is homogeneous, $H^*(\lambda z^0, t^0; a, c) = 0$ for all complex number λ. So $(H^*)^{-1}(0)$ is not a path in a neighborhood of (z^0, t^0), and thus 0 is not a regular value of $H^*(\cdot, \cdot; a, c)$ on $\{(z, t) : z \neq 0\}$. This is a contradiction. The proof is thus completed.¶

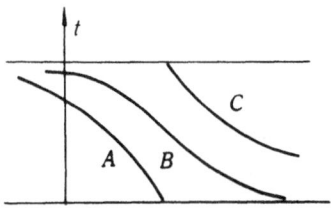

Figure 7.2

Theorem 1.13 asserts that almost every polynomial mapping has $q = \prod_{i=1}^{n} q_i$ distinct zeros. In other words, every polynomial mapping has q distinct zeros with probability 1. But for a specifically given polynomial mapping, the theorem has nothing to do with the number of the zeros of the mapping. The following Theorem 1.18 will give a sufficient condition that guarantees a special kind of polynomial mappings to have exactly q zeros.

For this purpose, we first introduce the concept of Brouwer degrees.

Definition 1.14 Let $D \subset \mathbf{R}^m$ be an open set and $F : D \to \mathbf{R}^m$ be a continuous mapping. Suppose that z^0 is an isolated zero of F. Let $D_1 \subset \overline{D}_1 \subset D$ be an open bounded neighborhood of z^0 such that $\overline{D}$ does not contain any other zero of F. Then call $\deg(F, \overline{D}_1, 0)$ the Brouwer degree of the zero z^0 of F.

We can verify that in the definition, the degree is independent of the choice of D_1. In fact, suppose that D_2 is another neighborhood of z^0 satisfying same conditions as D_1 and $\{F^k\}$ is a sequence of C^2 mapping which uniformly converges to F, and 0 is a regular value of all F^k. Let $D_3 \subset \overline{D}_3 \subset D_1 \cap D_2$ be an open neighborhood of z^0. Then the zeros of F^k in $D_1 \cup D_2$ are contained in D_3 as long as k large enough. Thus

$$
\begin{aligned}
\deg(F, \overline{D}_1, 0) &= \lim_{k \to \infty} \deg(F^k, \overline{D}_1, 0) \\
&= \lim_{k \to \infty} \deg(F^k, \overline{D}_3, 0) \\
&= \lim_{k \to \infty} \deg(F^k, \overline{D}_2, 0) \\
&= \deg(F, \overline{D}_2, 0).
\end{aligned}
$$

Definition 1.15 Let $P(\cdot; \omega) : \mathbf{C}^n \to \mathbf{C}^n$ be a polynomial mapping and z^0 be an isolated zero of $P(\cdot; \omega)$. Set D to be an open bounded neighborhood of z^0 which doesn't contain any other zeros of $P(\cdot; \omega)$. If

$P(\cdot; \omega)$ is regarded as a real self mapping of $\mathbf{R}^{2n}$ in the way as in Section 6.3, then the Brouwer degree $\deg(P, \overline{D}, 0) = \deg(P(\cdot; \omega), \overline{D}, 0) \geq 1$. (cf. the proof of Theorem 2.4). We call $\deg(P, \overline{D}, 0)$ the multiplicity of the zero z^0 of P.

From Theorem 1.7 we know that the multiplicity of a zero of a polynomial in one variable is consistent with the current definition of the generalized case.

Since $\det Q'(z) \neq 0$ at any zero of Q given by $Q_i(z) = z_i^{q_i} - 1$, $i = 1, \cdots, n$, the Brouwer degree or the multiplicity of any zero of Q is 1. Then with Theorem 2.11 in Chapter 6, we know that the multiplicity of any zero of almost every $P(\cdot; \omega)$ in Theorem 1.13 is also 1.

Lemma 1.16 *Given a polynomial mapping $P(z; \tilde{\omega})$. If $P^*(z; \tilde{a}, \tilde{c})$ has only the trivial zero $z = 0$ then $P(z; \tilde{\omega})$ has only a finite number of zeros.*

Proof. We first prove that the zero set of $P(z; \tilde{\omega})$ is bounded. Suppose otherwise that $\{z^k\}$ is a sequence of zeros of $P(z; \tilde{\omega})$ with

$$\lim_{k \to \infty} |z^k| = \infty.$$

Then as $k \to \infty$,

$$
\begin{aligned}
P_i^* \left(\frac{z^k}{|z^k|}; \tilde{a}, \tilde{c} \right) &= |z^k|^{-q_i} P_i^*(z^k; \tilde{a}, \tilde{c}) \\
&= |z^k|^{-q_i} [P_i^*(z^k; \tilde{a}, \tilde{c}) - P_i(z^k; \tilde{\omega})] \\
&\to 0.
\end{aligned}
$$

Thus, let z^0 be an accumulation point of $\{z^k/|z^k|\}$, we have

$$P^*(z^0; \tilde{a}, \tilde{c}) = 0 \text{ and } |z^0| = 1.$$

This contradicts to the hypotheses of the lemma. Hence, the zero set of $P(z; \tilde{\omega})$ is bounded. Then by Corollary 2.2 in [Rabinowitz, 1973], every zero of P is isolated. Therefore, $P(z; \tilde{\omega})$ has only a finite number of zeros.

¶

Lemma 1.17 *Let z^0 be an isolated zero of $P(z; \tilde{\omega}) = 0$ with multiplicity k. Suppose that D is an open bounded neighborhood of z^0 such that $\overline{D}$ doesn't contain any other zero of $P(z; \tilde{\omega})$. Then for all $e \in \mathbf{C}^k$ small enough such that 0 is a regular value of $P(\cdot; \tilde{\omega} + e)$, $P(z; \tilde{\omega} + e)$ has exactly k distinct zeros in D.*

Proof. Theorem 6.2.12 implies that for e small enough,

$$\deg(P(\cdot;\tilde{\omega}+e),\overline{D},0) = \deg(P(\cdot;\tilde{\omega}),\overline{D},0) = k.$$

Since 0 is a regular value of $P(\cdot;\tilde{\omega}+e)$, Theorem 6.3.2 and Remark 6.4.3 guarantee that as a real mapping on $\mathbf{R}^{2n}$, the Jacobian determinant of $P(\cdot;\tilde{\omega}+e)$ is always positive. Thus the number of zeros of $P(z;\tilde{\omega}+e)$ in D is $\deg(P(\cdot;\tilde{\omega}+e),\overline{D},0) = k.$ ¶

The next theorem was first discovered by Noether and Van der Waerden, see [Van der Waerden, 1931].

Theorem 1.18 *Given a polynomial mapping $P(z;\tilde{\omega})$. Let $P^*(z;\tilde{a},\tilde{c})$ be its corresponding homogeneous polynomial mapping with highest degree. If $P^*(z;\tilde{a},\tilde{c})$ has only the trivial zero $z = 0$ then $P(z;\tilde{\omega})$ has exactly $q = \prod_{i=1}^n q_i$ zeros, where q_i is the degree of P_i.*

Proof. By Theorem 1.13, for almost all e, $P(z;\tilde{\omega}+e)$ has exactly q zeros. Now, we assert that there is a positive number η such that for every e with $|e| \le \eta$, every zero of $P(z;\tilde{\omega}+e)$ satisfies $|z| \le M$ for some constant M. Suppose otherwise that there are $e^k \to 0$ and $|z^k| \to \infty$ such that $P(z^k;\tilde{\omega}+e^k) = 0$. Thus, as $e^k \to 0$,

$$
\begin{aligned}
P_i^*\left(\frac{z^k}{|z^k|};\tilde{a},\tilde{c}\right) &= |z^k|^{-q_i}[P_i^*(z^k;\tilde{a},\tilde{c}) - P_i(z^k;\tilde{\omega}+e^k)] \\
&= |z^k|^{-q_i}[P_i^*(z^k;\tilde{a},\tilde{c}) - P_i(z^k;\tilde{\omega}) - P_i(z^k;e^k)] \to 0.
\end{aligned}
$$

Let z^0 be an accumulation point of $\{z^k/|z^k|\}$. It is clear that

$$P^*(z^0;\tilde{a},\tilde{c}) = 0 \text{ and } |z^0| = 1.$$

This contradicts to that $P^*(z;\tilde{a},\tilde{c})$ has only the trivial zero.

Thus, we can take $\tilde{e}^k \to 0$ and $|\tilde{z}^k| \le M$ such that $P(\tilde{z}^k;\tilde{\omega}+e^k) = 0$. Let z^0 be an accumulation of $\{\tilde{z}^k\}$. Then $P(z^0;\tilde{\omega}) = 0$. This means that the zero set of $P(z^k;\tilde{\omega})$ is not empty. Besides, Lemma 1.16 shows that the set is finite. Let $\xi^1,\cdots,\xi^l$ be the all distinct zeros of $P(z^k;\tilde{\omega})$, and k_i is the multiplicity of ξ^i.

Take r small enough such that $\overline{D}_i \cap \overline{D}_j = \emptyset$ for $i \ne j$, where $D_i = \{z \in \mathbf{C}^n : |z-\xi^i| < r\}$, $i = 1,\cdots,l$. By Lemmas 1.11, 1.17 and the above boundedness of zeros, there is an e small enough such that $P(z;\tilde{\omega}+e)$ has exactly q distinct zeros which are contained in $\bigcup_{i=1}^l D_i$ and

$$\deg(P(\cdot;\tilde{\omega}+e),\overline{D}_i,0) = \deg(P(\cdot;\tilde{\omega}),\overline{D}_i,0) = k_i.$$

Thus, counting multiplicities, the number of zeros of $P(z; \tilde{\omega})$ is

$$
\sum_{i=1}^{l} k_i = \sum_{i=1}^{l} \deg(P(\cdot; \tilde{\omega} + e), \overline{D}_i, 0)
$$

$$
= q = \prod_{i=1}^{n} q_i. \P
$$

Remark 1.19 If the number of zeros of $P(z; \tilde{\omega})$ is not equal to q, counting multiplicities, then Theorem 1.18 implies that $P^*(z; \tilde{\omega})$ must have non-trivial zeros. For example,

$$
\begin{cases}
z_1^2 - z_2 - 1 = 0 \\
z_1^2 - 2z_2 = 0
\end{cases}
$$

has two simple zeros $(\sqrt{2}, 1)$ and $(-\sqrt{2}, 1)$. Thus, the $P_i^*(z; \tilde{\omega}) = z_1^2 = 0$, $i = 1, 2$, have a nontrivial zero $(0, r)$, $r \neq 0$.

Corollary 1.20 *Under the hypothesis of Theorem 1.18, if 0 is a regular value of $P(\cdot; \tilde{\omega})$ then the q zeros are all distinct.*

Proof. Notice that the multiplicity of every zero of $P(z; \tilde{\omega})$ is one since 0 is a regular value of $P(\cdot; \tilde{\omega})$. The result is now directly from Theorem 1.18. ¶

Theorem 1.21 (Bézout Theorem)   *Given a polynomial mapping $P(z; \tilde{\omega})$ with q_i the degree of its i-th component, $i = 1, \cdots, n$. Then the number of isolated zeros of $P(z; \tilde{\omega})$ is at most $q = \prod_{i=1}^{n} q_i$, counting multiplicities.*

Proof.   Let $\xi^1, \cdots, \xi^l$ be the all distinct isolated zeros of $P(z; \tilde{\omega})$ with multiplicities $k_1, \cdots, k_l$ respectively. Denote $D_i = \{z \in \mathbf{C}^n : |z - \xi^i| < r\}$. Choose r small enough such that $\overline{D}_i \cap \overline{D}_j = \emptyset$ for all $i \neq j$ and every $\overline{D}_i$ contains only one zero ξ_i, $i = 1, \cdots, l$. By Lemma 1.11, Theorem 1.13 and Lemma 1.17, there exists an $e \in \mathbf{C}^k$ small enough such that 0 is a regular value of $P(z; \tilde{\omega} + e)$, $P(z; \tilde{\omega} + e)$ has exactly $q = \prod_{i=1}^{n} q_i$ distinct zeros, and there are exactly k_i distinct zeros in each $\overline{D}_i$. Thus, counting multiplicities, the number of isolated zeros of $P(z; \tilde{\omega})$ is at most

$$
\sum_{i=1}^{l} k_i \leq q. \P
$$

Theorem 1.22 (Noether-Van der Waerden Theorem)   *If $P^*(z; \tilde{a}, \tilde{c})$ has only the trivial zero $z = 0$ then $P(z; \tilde{\omega})$ has at most $q = \prod_{i=1}^{n} q_i$ distinct zeros.*

Proof. This theorem is directly from Theorem 1.18, or from Lemma 1.16 and Theorem 1.21. ¶

Finally, we give some applications of Theorem 1.18.

Theorem 1.23 *Given a polynomial mapping $P(z; \tilde{\omega})$ with $P^*(z; \tilde{a}, \tilde{c})$ satisfying $P_i^*(z; \tilde{a}, \tilde{c}) = \prod_{j=1}^n e_{ij} z_j^r$, $i = 1, \cdots, n$, where all e_{ij} are complex numbers and r is a positive integer. If the matrix $e = (e_{ij})$ is nonsingular then $P(z; \tilde{\omega})$ has r^n zeros.*

Proof. Since $e = (e_{ij})$ is nonsingular, $P^*(z; \tilde{\omega}) = 0$, i.e., $\sum_{j=1}^n e_{ij} z_j^r = 0$, $i = 1, \cdots, n$ implies $(z_1^r, \cdots, z_n^r) = (0, \cdots, 0)$, that is, $(z_1, \cdots, z_n) = (0, \cdots, 0)$. Thus Theorem 1.18 gives the result. ¶

Theorem 1.24 *Given a polynomial mapping $P(z; \tilde{\omega})$ with $P^*(z; \tilde{a}, \tilde{c})$ of the form $P_i^*(z; \tilde{a}, \tilde{c}) = z_i^{s_i} \left(\sum_{j=1}^n e_{ij} z_j^r \right)$, $i = 1, \cdots, n$. If every principal submatrix of $e = (e_{ij})$ is nonsingular then $P(z; \tilde{\omega})$ has $\prod_{i=1}^n (r + s_i)$ zeros.*

Proof. Consider $z = (z_I, 0)$, $z_i \neq 0$ for $i \in I \subseteq \{1, \cdots, n\}$. If $P^*(z; \tilde{a}, \tilde{c}) = 0$ then $P_i^*(z; \tilde{a}, \tilde{c}) = z_i^{s_i} \left(\sum_{j=1}^n e_{ij} z_j^r \right) = 0$ for $i \in I$, that is,

$$\sum_{j \in I} e_{ij} z_j^r = 0, \quad i \in I.$$

Rewrite the above system of equations as $e_{II} y_I = 0$, where e_{II} is the principal submatrix of e formed by deleting i-row and i-column for $i \notin I$ and $y_i = z_i^r$ for $i \in I$. Since e_{II} is nonsingular, $y_I = 0$. Thus, $z_I = 0$. This contradiction gives the result. ¶

§2. Isolated Zeros

Let $P : \mathbf{C}^n \to \mathbf{C}^n$ be a given polynomial mapping, where $P = (P_1, \cdots, P_n)$ and the degree of P_i is $q_i \geq 1$. We have been familiar with the linear homotopy

$$H(z, t) = t P(z) + (1 - t) Q(z)$$

between the trivial polynomial mapping $H(z, 0) = Q(z)$ and the target polynomial mapping $H(z, 1) = P(z)$, where

$$Q_j(z) = z_j^{q_j} - b_j, \quad b_j \in \mathbf{C} - \{0\}, \quad 1 \leq j \leq n.$$

It is clear that $Q(z)$ has $q = \prod_{j=1}^n q_j$ distinct zeros. They are

$$(b_1^{1/q_1} e^{2k_1 \pi i / q_1}, \cdots, b_n^{1/q_n} e^{2k_n \pi i / q_n}), \quad k_j = 1, \cdots, q_j; \quad j = 1, \cdots, n.$$

By parameterized Sard's Theorem(1.10), one can choose a b such that 0 is a regular value of $H(\cdot, \cdot; b)$ in $\mathbf{C}^n \times [0, 1)$. Fix such a b. Then the zero set of H consists of some paths which may be unbounded as $t \to t_0 \leq 1$, that is, some paths may not approach to the zeros of $H(\cdot, 1; b)$, and thus some zeros of the mapping can't be found in this way.

To deal with this case, we introduce the term $t(1-t)R(z)$ into the homotopy, where

$$R_j(z) = \sum_{i=1}^n a_{ij} z_i^{q_j}, \quad a_{ij} \in \mathbf{C} \quad j = 1, \cdots, n.$$

It is clear that the term is zero when $t = 0$ or 1.

Consider the homotopy $H : \mathbf{C}^n \times [0, 1] \times \mathbf{C}^n \times \mathbf{C}^{n^2} \to \mathbf{C}^n$ defined by

$$H_j(z, t; b, a) = tP_j(z) + (1-t)(z_j^{q_j} - b_j) + t(1-t)\sum_{i=1}^n a_{ij} z_i^{q_j},$$

$j = 1, \cdots, n$, where $b \in \mathbf{C}^n$, $a = (a_{ij}) \in \mathbf{C}^{n^2}$ are parameters, and $t \in [0, 1]$ is the homotopy parameter.

Let $P^*(z) = (P_1^*(z), \cdots, P_n^*(z))$, where $P_j^* : \mathbf{C}^n \to \mathbf{C}$ consists of the terms of P_j with degree q_j. Then we obtain $H^* : \mathbf{C}^n \times [0, 1] \times \mathbf{C}^{n^2} \to \mathbf{C}^n$ defined by

$$H_j^*(z, t; a) = tP_j^*(z) + (1-t)z_j^{q_j} + t(1-t)\sum_{i=1}^n a_{ij} z_i^{q_j},$$

$j = 1, \cdots, n$.

By Parameterized Sard's Theorem(1.10), we have

Lemma 2.1   *For almost all $(b, a) \in \mathbf{C}^n \times \mathbf{C}^{n^2}$, $0 \in \mathbf{C}^n$ is a regular value of $H(\cdot, \cdot; b, a)$ in $\mathbf{C}^n \times [0, 1)$ and is also a regular value of $H^*(\cdot, \cdot; a)$ in $(\mathbf{C}^n - \{0\}) \times [0, 1)$.*

Proof.   It is clear that for $t \in [0, 1)$

$$\operatorname{rank} H_b' = \operatorname{rank}(1-t) \begin{pmatrix} -1 & & \\ & \ddots & \\ & & -1 \end{pmatrix} = n.$$

Then Theorem 1.10 implies that for almost all $(b, a) \in \mathbf{C}^n \times \mathbf{C}^{n^2}$, 0 is a regular value of $H(\cdot, \cdot; b, a)$.

If $t \neq 0, 1$ and $z \neq 0$. Choose an i such that $z_i \neq 0$. Then the derivatives of H^* with respect to $a_{i \cdot} = (a_{i1}, \cdots, a_{in})$ are

$$
H_{a_{i \cdot}}^{*'} = \begin{pmatrix} t(1-t)z_i^{q_1} & & \\ & \ddots & \\ & & t(1-t)z_i^{q_n} \end{pmatrix}.
$$

Thus, $H_{a_{i \cdot}}^{*'}$ is of full rank. If $t = 0$, $H^*(z, 0; a) = (z_1^{q_1}, \cdots, z_n^{q_n}) \neq 0$. In summary, by Theorem 1.10, for almost all $a \in \mathbf{C}^{n^2}$, 0 is a regular value of $H^*(\cdot, \cdot; a)$ in $(\mathbf{C}^n - \{0\}) \times [0, 1)$. This finishes the proof. ¶

Lemma 2.2 *Fix an a as in Lemma 2.1 such that 0 is a regular value of $H^*(\cdot, \cdot; a)$ in $(\mathbf{C}^n - \{0\}) \times [0, 1])$. If $H^*(z^0, t_0; a) = 0$ and $t_0 \in [0, 1)$, then $z^0 = 0$.*

Proof. Suppose otherwise that $z^0 \neq 0$. Since 0 is a regular value of $H^*(\cdot, \cdot; a)$ in $(\mathbf{C}^n - \{0\}) \times [0, 1)$, there is a real one-dimensional curve in $(H^*)^{-1}(0) = \{(z, t) \in \mathbf{C}^n \times [0, 1] \times \mathbf{C}^{n^2} : H^*(z, t; a) = 0\}$ running through (z^0, t_0). But since H^* is homogeneous in z, $H^*(\lambda z^0, t_0; a) = 0$ for any $\lambda \in \mathbf{C} = \mathbf{R}^2$. This means that $(H^*)^{-1}(0)$ contains a real two-dimensional plane. This contradiction gives the result. ¶

Lemma 2.3 *Let (b, a) be as in Lemma 2.1 such that 0 is a regular value of H and of $H^*(\cdot, \cdot; a)$ in $(\mathbf{C}^n - \{0\}) \times [0, 1])$. Then for any fixed $t_0 \in [0, 1)$ there is a positive constant $K(t_0)$ such that for every zero (z, t) of H in $\mathbf{C}^n \times [0, t_0]$, $|z| \leq K(t_0)$.*

Proof. Suppose otherwise that there are some $t_0 \in [0, 1)$ and a sequence of $(z^k, t_k) \in \mathbf{C}^n \times [0, t_0]$ such that $H(z^k, t_k; b, a) = 0$ and $|z^k| \to \infty$. With on loss of generality, assume $z^k / |z^k| \to z^0 \in \mathbf{C}^n$ and $t_k \to \tau \in [0, t_0]$.

Notice that the degree of $H_j - H_j^*$ in z is less than q_j, we have that as $k \to \infty$,

$$
H_j^* \left(\frac{z^k}{|z^k|}, t_k; a \right) = |z^k|^{-q_j} H_j^*(z^k, t_k; a)
$$

$$
= |z^k|^{-q_j} [H_j^*(z^k, t_k; a) - H_j(z^k, t_k; b, a)]
$$

$$
\to 0.
$$

Thus, $H^*(z^0, \tau; a) = 0$ with $|z^0| = 1$. This contradicts to Lemma 2.2. The proof is thus completed. ¶

Now, we are ready to prove our main result of this section.

Theorem 2.4 *Let $P : \mathbf{C}^n \to \mathbf{C}^n$ be a polynomial mapping and the positive integer q_j be the degree of its j-th component, $j = 1, \cdots, n$. Consider the homotopy*

$$H : \mathbf{C}^n \times [0,1] \times \mathbf{C}^n \times \mathbf{C}^{n^2} \to \mathbf{C}^n,$$

its j-th component is

$$H_j(z, t; b, a) = tP_j(z) + (1-t)(z_j^{q_j} - b_j) + t(1-t)\sum_{i=1}^{n} a_{ij} z_i^{q_j},$$

where $b \in \mathbf{C}^n$, $a \in \mathbf{C}^{n^2}$, $t \in [0,1]$. Then for almost all $(b, a) \in \mathbf{C}^n \times \mathbf{C}^{n^2}$, the zero set

$$\{(z, t) \in \mathbf{C}^n \times [0,1) : H(z, t; b, a) = 0\}$$

consists of $q = \prod_{j=1}^{n} q_j$ distinct analytic paths $z^1(t), \cdots, z^q(t)$. These paths respectively start from the q distinct zeros of $H(\cdot, 0; b, a)$. If every path is 1 by its arc length s then $\dfrac{dt}{ds} > 0$.

For every j, we have either

$$\lim_{\substack{l \to \infty \\ t \to 1^-}} |z^j(t_l)| = \infty$$

or the set $\{z^{j*} : \lim_{\substack{l \to \infty \\ t \to 1^-}} z^j(t_l) = z^{j*}\}$ forming a connected subset of $P^{-1}(0)$.

Moreover, for every isolated zero z^0 of P, there exists some j such that

$$\lim_{t \to 1^-} z^j(t) = z^0.$$

In particular, $P(z)$ has at most q isolated zeros. (It is the classical Bézout Theorem, and here we provide an approach of homotopy methods)

Proof. By Lemma 2.1, for almost every $(b, a) \in \mathbf{C}^n \times \mathbf{C}^{n^2}$, 0 is a regular value of $H(\cdot, \cdot; b, a)$ in $\mathbf{C}^n \times [0,1)$. Thus the zero set

$$H^{-1}(0) = \{(z, t) \in \mathbf{C}^n \times [0,1) : H(z, t; b, a) = 0\}$$

consists of $q = \prod_{j=1}^{n} q_j$ distinct analytic paths $z^1(t), \cdots, z^q(t)$ which start from the q distinct zeros of $H(\cdot, 0; b, a)$.

By Lemma 2.3, the above paths may be unbounded only when $t \to 1^-$. With Remark 6.4.3, it is clear that $\dfrac{dt}{ds} > 0$. It is also easy to see that $\{z^{j*} : \lim_{\substack{l \to \infty \\ t \to 1^-}} z^j(t_l) = z^{j*}\}$ is a connected subset of $P^{-1}(0)$.

Finally, we will prove that for every isolated zero z^0 of P there is some j such that $\lim_{t \to 1^-} z^j(t) = z^0$. For this, it suffices to prove that $H(\cdot, t; b, a)$ has zeros approaching z^0 as $t \to 1^-$.

Let $U(r) = \{z : |z - z^0| < r\}$. Choose r small enough such that $\overline{U(r)}$ doesn't contain any isolated zero of P other than z^0. Then Theorem 6.2.12 gives that there exists a positive number δ such that if $1 - \delta < t < 1$ then

$$\deg(H(\cdot, t; b, a), \overline{U(r)}, 0) = \deg(P, \overline{U(r)}, 0).$$

Let

$$\mu(x, y) = f(x, y)dx \wedge dy = f(x, y)dx_1 \wedge \cdots \wedge dx_n \wedge dy_1 \wedge \cdots \wedge dy_n$$

represent the C^∞ $2n$-form in $\mathbf{R}^{2n}$ as shown in Theorem 6.2.17. In particular, we can choose f such that $f(x, y) \geq 0$, and $f(x, y) > 0$ near the origin.

By Lemma 1.11, Theorems 1.13 and 6.2.12, we can choose a polynomial mapping $\tilde{P} : \mathbf{C}^n \to \mathbf{C}^n$ such that

$$\deg(\tilde{P}, \overline{U(r)}, 0) = \deg(P, \overline{U(r)}, 0),$$

$f \circ \tilde{P}(z^0) > 0$, $\tilde{P}$ has exactly q distinct zeros and 0 is a regular value of $\tilde{P}$. It is clear that $\det \tilde{P}'(z) \not\equiv 0$. Thus $\det \tilde{P}'(z) \not\equiv 0$ on any open set of $\mathbf{C}^n$. In summary,

$$
\begin{aligned}
\deg(H(\cdot, t; b, a), \overline{U(r)}, 0) &= \deg(P, \overline{U(r)}, 0) \\
&= \deg(\tilde{P}, \overline{U(r)}, 0) \\
&= \int_{U(r)} f \circ \tilde{P}(z) |\det \tilde{P}'(z)|^2 dx \wedge dy \\
&> 0.
\end{aligned}
$$

Notice that the degree is always an integer, this leads to

$$\deg(H(\cdot, t; b, a), \overline{U(r)}, 0) = \deg(P, \overline{U(r)}, 0) \geq 1.$$

Then by Theorem 6.2.14, $H(\cdot, t; b, a)$ has at least one zero in $U(r)$.

It is easy to show that every path of $z^1(t), \cdots, z^q(t)$ approaches at most one isolated zero of P. Thus P has at most q isolated zeros. Furthermore, if z^0 is an isolated zero with multiplicity k then there are exactly k paths which tend to z^0. ¶

The homotopy H in Theorem 2.4 was introduced by Chow, Mallet-Paret and Yorke (see [Chow, Mallet-Paret & Yorke, 1978]). T. Y. Li presented a fairly simple homotopy $H : \mathbf{C}^n \times [0.1] \times \mathbf{C}^n \times \mathbf{C}^n \to \mathbf{C}^n$ defined by

$$H(z, t; a, b) = tP(z) + (1 - t)Q(z),$$

where $Q(z) = (Q_1(z), \cdots, Q_n(z)) \in \mathbf{C}^n$ and

$$Q_k(z) = a_k z_k^{q_k} - b_k, \quad a_k, b_k \in \mathbf{C}, 1 \le k \le n.$$

The results of Theorem 2.4 are also true for this homotopy.

By Theorem 2.4 and Bézout Theorem, we now reprove Theorem 1.8. The proof itself is also interesting.

Theorem 2.5 *Let P_1 and P_2 be two complex polynomials with no common factors. The degree of P_j is q_j, $j = 1, 2$. Then the number of the solutions of the system of equations*

$$\begin{cases} P_1(z_1, z_2) = 0 \\ P_2(z_1, z_2) = 0 \end{cases}$$

is at most $q = q_1 q_2$, counting multiplicities.

Proof. Denote $q_{z_1}(P_j)$ the degree of P_j in z_1. Let

$$\begin{aligned} P_1 &= d_1 P_2 + P_3, & q_{z_1}(P_3) &< q_{z_1}(P_2), \\ P_2 &= d_2 P_3 + P_4, & q_{z_1}(P_4) &< q_{z_1}(P_3), \end{aligned}$$

$$\cdots$$

$$P_{r-1} = d_{r-1} P_r + P_{r+1}, \quad q_{z_1}(P_{r+1}) < q_{z_1}(P_r),$$

where $q_{z_1}(P_{r+1}) = 0$. It is clear that

$$P_{r+1} = P_{r+1}(z_2) = \frac{\tilde{P}_{r+1}(z_2)}{\tilde{\tilde{P}}_{r+1}(z_2)},$$

where $\tilde{P}_{r+1}$ and $\tilde{\tilde{P}}_{r+1}$ are polynomials in z_2.

Suppose that the system of equations

$$\begin{cases} P_1(z_1, z_2) = 0 \\ P_2(z_1, z_2) = 0 \end{cases}$$

has infinitely many solutions (ξ_k, η_k), $k = 1, 2, \cdots$. Notice that P_1 and P_2 have no common factors, with no loss of generality, we assume that $\xi_k \neq \xi_l$ and $\eta_k \neq \eta_l$ for $k \neq l$, and every η_k, $k = 1, 2, \cdots$, is such that all coefficients of d_i and P_i in z_1 are well defined, that is, the denominators of the concerned rational functions in z_2 are not zero at η_k. It is clear that $P_1(\xi_k, \eta_k) = 0, \cdots, P_r(\xi_k, \eta_k) = 0$ and $\tilde{P}_{r+1}(\eta_k) = 0$, $\tilde{\tilde{P}}_{r+1}(\eta_k) = 0$. Since $\tilde{P}_{r+1}(\eta_k) = 0$, $k = 1, 2, \cdots$ imply that $\tilde{P}_{r+1} \equiv 0$, we have $P_{r+1} \equiv 0$.

Notice that $P_r(z_1, z_2) = \tilde{f}_r(z_1, z_2)/\tilde{\tilde{f}}_r(z_2)$, where $\tilde{f}_r$ and $\tilde{\tilde{f}}_r$ have no common factors and $q_{z_1}(\tilde{P}_r) \geq 1$. Then every factor of $\tilde{P}_r$ must be a common factor of P_1 and P_2. This contradicts to the hypothesis of the theorem.

Therefore, the number of the solutions of the system of equations is finite and every solution of the system is isolated. Finally, Theorem 1.21 or Theorem 2.4 guarantees that the number is at most $q = q_1 q_2$. ¶

Corollary 2.6 *Let F be an irreducible polynomial in x and y and its degree be $\delta > 0$. Then F has at most $(\delta - 1)^2$ singular points.*

Proof. Notice that singular points of F are the solutions of the system of equations

$$\begin{cases} F(x, y) = 0, \\ \dfrac{\partial F}{\partial x}(x, y) = 0, \\ \dfrac{\partial F}{\partial y}(x, y) = 0. \end{cases}$$

If $\dfrac{\partial F}{\partial x} = h \cdot F_1$, $\dfrac{\partial F}{\partial y} = h \cdot F_2$, where the degree of h is at most $\delta - 2$, h, F_1 and F_2 are polynomials, and F_1 and F_2 have no common factors, then the above system is equivalent to

$$\begin{cases} F = 0 \\ h = 0 \end{cases} \quad \text{and} \quad \begin{cases} F = 0 \\ F_1 = 0 \\ F_2 = 0. \end{cases}$$

Since F is irreducible and the degree of h is less than that of F, F and h have no common factors. Then Theorem 2.5 gives that the number of singular points is at most

$$\begin{aligned} (\delta - 1 - k)^2 + \delta k &= (\delta - 1)^2 - 2k(\delta - 1) + k^2 + \delta k \\ &= (\delta - 1)^2 - k(\delta - 2 - k) \end{aligned}$$

$$\leq \ (\delta - 1)^2.$$

If $\dfrac{\partial F}{\partial y} \not\equiv 0$ and $\dfrac{\partial F}{\partial x} \equiv c \dfrac{\partial F}{\partial y}$, where c is a constant. Notice that now $F(x, y)$ has infinitely many 1 while

$$\begin{cases} F(x, y) = 0 \\ \dfrac{\partial F}{\partial y}(x, y) = 0 \end{cases}$$

has only a finite number of solutions since F and $\dfrac{\partial F}{\partial x}$ have no common factors. Thus there must exist (x_0, y_0) such that

$$\begin{cases} F(x_0, y_0) = 0 \\ \dfrac{\partial F}{\partial y}(x_0, y_0) \neq 0. \end{cases}$$

By Implicit Function Theorem, there is an analytic function $y = y(x)$ near x_0 such that $y(x_0) = y_0$. Thus,

$$\frac{\partial F}{\partial x} + \frac{\partial F}{\partial y}\frac{dy}{dx} \equiv \frac{\partial F}{\partial y}\left(c + \frac{dy}{dx}\right) = 0,$$

$$\frac{dy}{dx} = -c, \quad y = -cx + d.$$

Rewrite $F(x, y)$ as

$$F(x, y) = a_k(x)(y + cx - d)^k + a_{k-1}(x)(y + cx - d)^{k-1} + \cdots + a_0(x).$$

Then near x_0,

$$0 \equiv F(x, -cx + d) \equiv a_0(x).$$

This leads to $a_0(x) \equiv 0$, and thus the irreducible polynomial F has the factor $y + cx - d$. Therefore, $F(x, y) \equiv c_1(y + cx - d)$, where c_1 is a constant. It is clear that F has no singular points.

If $\dfrac{\partial F}{\partial y} \equiv 0$ then $F(x, y) = f(x)$. Thus

$$\begin{cases} f(x) = 0 \\ \dfrac{\partial f}{\partial x} = 0 \end{cases}$$

is solvable if and only if $f(x)$ has multiple zeros. But f is irreducible. So it has no singular points. ¶

Remark 2.7 Corollary 2.6 can also be obtained from Theorem 4.4 of [Walker, 1950].

§3 Locating Zeros of Analytic Functions in Bounded Regions

This section concerns with the problems of finding the zeros of analytic functions in a bounded region.

Lemma 3.1 *Let f be an analytic function in the plane $\mathbf{C}$. Also let L be a simple closed curve on $\mathbf{C}$ which doesn't contain the zeros of f and D be its interior. Denote*

$$s_k = \frac{1}{2\pi i} \int_L z^k \frac{f'(z)}{f(z)} dz.$$

Then

$$s_k = \sum_{j=1}^{n} \xi_j^k,$$

where ξ_j, $j = 1, 2 \cdots, n$, are the all zeros of f in D, counting multiplicities.

Proof. The proof of this lemma is similar to that of Lemma 1.3 and is thus omitted. ¶

Remark 3.2 Given an analytic function f, we can calculate s_k, $k = 1, \cdots, n$ by the formula in Lemma 3.1. Then to find the zeros of f in D, we need only to find one zero of the following system of equations:

$$(*) \quad \begin{cases} s_1 = \xi_1 + \xi_2 + \cdots + \xi_n \\ s_2 = \xi_1^2 + \xi_2^2 + \cdots + \xi_n^2 \\ \cdots \\ s_n = \xi_1^n + \xi_2^n + \cdots + \xi_n^n. \end{cases}$$

Next, we will only concern with the system $(*)$. We first introduce the following Newton's identities.

Lemma 3.3 *Let*

$$\begin{cases} \sigma_1 = -(\xi_1 + \xi_2 + \cdots + \xi_n) \\ \sigma_2 = \xi_1 \xi_2 + \cdots + \xi_{n-1}\xi_n \\ \cdots \\ \sigma_n = (-1)^n \xi_1 \cdots \xi_n. \end{cases}$$

Then the Newton's identities are

$$\begin{cases} s_1 + \sigma_1 = 0 \\ s_2 + s_1\sigma_1 + 2\sigma_2 = 0 \\ s_3 + s_2\sigma_1 + s_1\sigma_2 + 3\sigma_3 = 0 \\ \cdots \\ s_n + s_{n-1}\sigma_1 + \sigma_{n-2}\sigma_2 + \cdots + s_1\sigma_{n-1} + n\sigma_n = 0. \end{cases}$$

Proof. Since all $\xi_j, j = 1, \cdots, n$ are the zeros of the polynomial

$$\prod_{j=1}^{n} (z - \xi_j) = z^n + \sigma_1 z^{n-1} + \sigma_2 z^{n-2} + \cdots + \sigma_n,$$

we have

$$s_n + s_{n-1}\sigma_1 + \sigma_{n-2}\sigma_2 + \cdots + s_1\sigma_{n-1} + n\sigma_n$$

$$= \sum_{j=1}^{n} (\xi_j^n + \sigma_1\xi_j^{n-1} + \sigma_2\xi_j^{n-2} + \cdots + \sigma_n) = 0.$$

Now, we prove inductively the identities.

The case of $n = 1$ is obvious. Suppose that the identities are true for $n = m - 1$. Now we prove the identities is also true for $n = m$. In fact, for $k \leq m - 1$, let

$$\bar{s}_j = \xi_1^j + \cdots + \xi_{m-1}^j,$$

$$\bar{\sigma}_j = (-1)^j (\xi_1 \cdots \xi_j + \cdots + \xi_{m-j} \cdots \xi_{m-1}).$$

Then by the induction hypothesis,

$$s_k + s_{k-1}\sigma_1 + s_{k-2}\sigma_2 + \cdots + k\sigma_k$$

$$= (\bar{s}_k + \xi_m^k) + (\bar{s}_{k-1} + \xi_m^{k-1})(\bar{\sigma}_1 - \xi_m)$$

$$+ (\bar{s}_{k-2} + \xi_m^{k-2})(\bar{\sigma}_2 - \bar{\sigma}_1\xi_m) + (\bar{s}_{k-3} + \xi_m^{k-3})(\bar{\sigma}_3 - \bar{\sigma}_2\xi_m)$$

$$+ \cdots$$

$$+ (\bar{s}_1 + \xi_m)(\bar{\sigma}_{k-1} - \bar{\sigma}_{k-2}\xi_m) + [(k-1) + 1](\bar{\sigma}_k - \bar{\sigma}_{k-1}\xi_m)$$

$$= (\bar{s}_k + \bar{s}_{k-1}\bar{\sigma}_1 + \bar{s}_{k-2}\bar{\sigma}_2 + \cdots + \bar{s}_1\bar{\sigma}_{k-1} + (k-1)\bar{\sigma}_k)$$

$$- \xi_m (\bar{s}_{k-1} + \bar{s}_{k-2}\bar{\sigma}_1 + \bar{s}_{k-3}\bar{\sigma}_2 + \cdots + \bar{s}_1\bar{\sigma}_{k-2} + (k-1)\bar{\sigma}_{k-1})$$

$$= 0 + 0 = 0.$$

Moreover, it is clear that

$$s_m + s_{m-1}\sigma_1 + s_{m-2}\sigma_2 + \cdots + s_1\sigma_{m-1} + m\sigma_m = 0.\P$$

Remark 3.4 From Lemma 3.3 we know that σ_j, $j = 1, \cdots, n$ can be uniquely determined from $s_i, \cdots, s_n$ and vice versa. Furthermore, every component of the solutions of the system $(*)$ is a zero of the polynomial

$$F(z) = z^n + \sigma_1 z^{n-1} + \cdots + \sigma_n.$$

Due to the highly sensitive dependence of a high degree polynomial on its coefficients, we don't intend to solve the polynomial F for the zeros of f.

Next, we will provide a stable method which solves the polynomial system $(*)$ directly without forming the polynomial F.

Lemma 3.5 Let $P : \mathbf{C}^n \to \mathbf{C}^n$ be a polynomial mapping defined by

$$P_k(z) = z_1^k + \cdots + z_n^k, \quad k = 1, \cdots, n.$$

Then 0 is the unique zero of P.

Proof. It is clear that $P(0) = 0$. On the other hand, suppose that $P(z) = 0$, that is, $s = (s_1, \cdots, s_n) = (0, \cdots, 0)$. By the Newton's identities, we have $(\sigma_1, \cdots, \sigma_n) = (0, \cdots, 0)$ and for $j = 1, \cdots, n$,

$$0 = z_j^n + z_j^{n-1}\sigma_1 + \cdots + z_j\sigma_{n-1} + \sigma_n = z_j^n.$$

Thus, $z = (z_1, \cdots, z_n) = (0, \cdots, 0)$. ¶

In the rest of this section, the P concerned is always the one as defined in Lemma 3.5, and $s = (s_1, \cdots, s_n)$ is given.

Theorem 3.6

(1) If $\xi = (\xi_1, \cdots, \xi_n)$ is a solution of the polynomial mappings $P(z) = s$ then $\xi_\mu = (\xi_{\mu(1)}, \cdots, \xi_{\mu(n)})$ is also a solution of $P(z) = s$ for every permutation μ of $(1, \cdots, n)$.

(2) If $\xi = (\xi_1, \cdots, \xi_n)$ and $\tilde{\xi} = (\tilde{\xi}_1, \cdots, \tilde{\xi}_n)$ are two solutions of $P(z) = s$ then there is a permutation μ of $(1, \cdots, n)$ such that $\tilde{\xi} = (\tilde{\xi}_1, \cdots, \tilde{\xi}_n) = (\xi_{\mu(1)}, \cdots, \xi_{\mu(n)})$.

(3) $P(z) = s$ has $n!$ solutions, counting multiplicities.

Proof.

(1) Since $P(z) = s$ is symmetric in z and $\xi = (\xi_1, \cdots, \xi_n)$ is a solution of $P(z) = s$, it is easy to show that $\xi_\mu = (\xi_{\mu(1)}, \cdots, \xi_{\mu(n)})$ is also a solution of $P(z) = s$.

(2) Let

$$\sigma_j = (-1)^j (\xi_1 \cdots \xi_j + \cdots + \xi_{n-j+1} \cdots \xi_n),$$
$$\tilde{\sigma}_j = (-1)^j (\tilde{\xi}_1 \cdots \tilde{\xi}_j + \cdots + \tilde{\xi}_{n-j+1} \cdots \tilde{\xi}_n).$$

Notice that $P(\tilde{\xi}) = s = P(\xi)$, we have $\sigma_j = \tilde{\sigma}_j$ for $j = 1, \cdots, n$, and thus

$$
\prod_{j=1}^{n} (z - \tilde{\xi}_j) = z^n + z^{n-1}\tilde{\sigma}_1 + \cdots + \tilde{\sigma}_n
$$
$$
= z^n + z^{n-1}\sigma_1 + \cdots + \sigma_n
$$
$$
= \prod_{j=1}^{n} (z - \xi_j).
$$

This implies that $(\tilde{\xi}_1, \cdots, \tilde{\xi}_n) = (\xi_{\mu(1)}, \cdots, \xi_{\mu(n)})$, where μ is a permutation of $(1, \cdots, n)$.

(3) By Lemma 3.5, $P^*(z) = P(z) = 0$ has a unique solution $z = 0$. Then Theorem 1.18 gives that $P(z) = s$ has exactly $n!$ solutions, counting multiplicities. ¶

Let $U = \{(z_1, \cdots, z_n) \in \mathbf{C}^n : z_i \neq z_j \text{ for } i \neq j\}$. It is clear that U is an open set of $\mathbf{C}^n$.

Lemma 3.7   *For every* $z \in U$, $P'(z)$ *is nonsingular.*

Proof.   This is just a Vandermonde determinant argument:

$$
\det P'(z) = \begin{vmatrix} 1 & 1 & \cdots & 1 \\ 2z_1 & 2z_2 & \cdots & 2z_4 \\ \cdots & & & \\ nz_1^{n-1} & nz_2^{n-1} & \cdots & nz_n^{n-1} \end{vmatrix}
$$
$$
= n! \begin{vmatrix} 1 & 1 & \cdots & 1 \\ z_1 & z_2 & \cdots & z_4 \\ \cdots & & & \\ z_1^{n-1} & z_2^{n-1} & \cdots & z_n^{n-1} \end{vmatrix}
$$
$$
= n! \prod_{1 \leq i < j < n} (z_j - z_i) \neq 0, \quad \forall z \in U. ¶
$$

Consider the homotopy $H : \mathbf{C}^n \times (0,1) \times U \to \mathbf{C}^n$ defined by

$$H(z, t; a) = t(P(z) - s) + (1 - t)(P(z) - P(a)).$$

It is easy to know that a is a solution of $H(z, 0; a) = P(z) - P(a) = 0$ and $H(z, 1; a) = P(z) - s$.

Next theorem gives the fundamental of the homotopy method of finding the solutions of $P(z) = s$.

Theorem 3.8 (1) 0 *is a regular value of* $H(\cdot, \cdot; \cdot)$.

(2) *For almost every fixed* $a \in U$, *the component of*

$$H_a^{-1}(0) = \{(z, t) \in \mathbf{C}^n \times (0, 1) :$$
$$H(z, t; a) = t(P(z) - s) + (1 - t)(P(z) - P(a)) = 0\}$$

with $(a_\mu, 0)$ *as a boundary point is a smooth curve which has the other boundary point* $(z_\mu^0, 1)$ *with* $P(z_\mu^0) = s$, *where* $a_\mu = (a_{\mu(1)}, \cdots, a_{\mu(n)})$ *and* μ *is a permutation of* $(1, \cdots, n)$.

Proof.

(1) Suppose that for some $(\tilde{z}, \tilde{t}; \tilde{a}) \in \mathbf{C}^n \times (0, 1) \times U$, $H(\tilde{z}, \tilde{t}; \tilde{a}) = 0$. Then the Jacobian matrix $H_a'(\tilde{z}, \tilde{t}; \tilde{a}) = -(1 - \tilde{t})P'(\tilde{a})$. Lemma 3.7 guarantees that for $\tilde{a} \in U$ and $\tilde{t} \neq 1$, the Jacobian matrix is nonsingular. Thus, regarding H as a real homotopy, we have

$$2n \geq \text{rank}_{\mathbf{R}} H'(\tilde{z}, \tilde{t}; \tilde{a}) \geq \text{rank}_{\mathbf{R}} H_a'(\tilde{z}, \tilde{t}; \tilde{a}) = 2n.$$

This means that $\text{rank}_{\mathbf{R}} H_a'(\tilde{z}, \tilde{t}; \tilde{a}) = 2n$, and thus 0 is a regular value of $H(\cdot, \cdot; \cdot)$.

(2) By parameterized Sard's Theorem(1.10), for almost every $a \in U$, 0 is a regular value of $H_a(z, t) \equiv H(z, t; a)$. Then Implicit Function Theorem gives that every component of $H_a^{-1}(0)$ is diffeomorphic either to a circle or to an open interval. But Remark 6.4.3 guarantees that the component is not diffeomorphic to a circle. Thus the component of $H_a^{-1}(0)$ starting from $(a_\mu, 0)$ will converge either to ∞ or to $(z_\mu^0, 1)$, where $P(z_\mu^0) = s$. Now we only need to prove that $H_a^{-1}(0)$ is bounded.

Suppose otherwise that there is a sequence $\{(z^k, t_k)\} \subset H^{-1}(0)$ with $|z^k| \to \infty$. Consider $P_j(z^k / |z^k|)$, $j = 1, \cdots, n$. Since P_j is a homogeneous polynomial of degree j, we have

$$P_j \left(\frac{z^k}{|z^k|} \right) = \frac{1}{|z^k|^j} P_j(z^k)$$

$$= \frac{1}{|z^k|^j} (t_k s + (1 - t_k)P(a))$$

$$\to 0, \quad \text{as } k \to \infty.$$

Let $\tilde{z}$ be an accumulation point of $\{z^k/|z^k|\}$. Then $P(\tilde{z}) = 0$ and $|\tilde{z}| = 1$. This contradicts to Lemma 3.5. ¶

Choose an $a \in U$ such that 0 is a regular value of $H_a(z,t)$. Then $H_a^{-1}(0)$ has exactly $n!$ distinct smooth paths and by Remark 6.4.3, t is nondecreasing along every path. By Remark 3.2, we need only to follow one path in the discussion of finding the all zeros of the analytic function f in the domain D.

Let $\Gamma_{a_\mu} = \{(z(\theta), t(\theta)) : 0 \le \theta < \theta_1\}$ is a path in $H_a^{-1}(0)$ with $(z(0), t(0)) = (a_\mu, 0)$. We have that, by $H_a(z(\theta), t(\theta)) = 0$, Γ_{a_μ} is the unique solution of

$$(**) \qquad \begin{cases} (D_z H_a)\dfrac{dz}{d\theta} + (D_t H_a)\dfrac{dt}{d\theta} = 0, \\[2mm] (z(0), t(0)) = (a_\mu, 0), \end{cases}$$

where $D_z H_a = P'(z)$, the Jacobian matrix of H_a with respect to z, and $D_t H_a = P(a) - s$. First, we prove $\dfrac{dt}{d\theta} \neq 0$. Suppose otherwise that $\dfrac{dt}{d\theta} = 0$ for some $(\tilde{z}, \tilde{t}) \in \Gamma_{a_\mu}$. Then $(D_z H_a)\dfrac{dz}{d\theta} = 0$. Since $\dfrac{dz}{d\theta} \neq 0$, $D_z H_a(\tilde{z}, \tilde{t}) = P'(\tilde{z})$ is singular. This contradicts Lemma 3.7. Moreover, since $t \ge 0$ and $t(0) = 0$, we have $\dfrac{dt}{d\theta} > 0$, and thus t is strictly increasing in θ. Then we may rewrite $(**)$ as

$$\begin{cases} \dfrac{dz}{dt} = -(D_z H_a)^{-1}(D_t H_a) = -(P'(z))^{-1}(P(a) - s) \\[2mm] z(0) = a_\mu. \end{cases}$$

The rest of this section provides a convenient way to calculate the inverse of the matrix $P'(z)$.

Let $A = (a_{ij})$ be an $n \times n$ matrix. Recall that the minor M_{ij} of a_{ij} is the $(n-1) \times (n-1)$ submatrix with the i-th row and the j-th column of A deleted. The cofactor A_{ij} of a_{ij} is defined to be $(-1)^{i+j} \det M_{ij}$. If the inverse of A is $(\hat{A}_{ij})$, then

$$\hat{A}_{ij} = \frac{1}{\det A} A_{ji}.$$

For any $i = 1, \cdots, n$, let $\sigma_1^i, \cdots, \sigma_{n-1}^i$ be respectively defined as

$$
\begin{cases}
\sigma_1 = -(z_1 + \cdots + z_n) \\
\sigma_2 = z_1 z_2 + \cdots + z_{n-1} z_n \\
\cdots \\
\sigma_n = (-1)^n z_1 \cdots z_n
\end{cases}
$$

with $z_i = 0$. For example, if $i = 1$ then

$$
\begin{cases}
\sigma_1^1 = -(z_2 + \cdots + z_n) \\
\sigma_2^1 = z_2 z_3 + \cdots + z_{n-1} z_n \\
\cdots \\
\sigma_{n-1}^1 = (-1)^{n-1} z_2 \cdots z_n
\end{cases}
$$

Besides, we set $\sigma_0^i = 1$.

Lemma 3.9 *Let A be an $n \times n$ Vandermonde matrix, i.e.,*

$$
A = (a_{ij}) = (z_j^{i-1}) =
\begin{pmatrix}
1 & \cdots & 1 \\
z_1 & \cdots & z_n \\
\cdots & & \\
z_1^{n-1} & \cdots & z_n^{n-1}
\end{pmatrix}.
$$

Then the cofactor of a_{ij} is

$$
A_{ij} = (-1)^{n+j} \prod_{\substack{k > k' \\ k \neq j, k' \neq j}} (z_k - z_{k'}) \sigma_{n-i}^j.
$$

Proof. It is clear that

$$
A_{ij} = (-1)^{i+j}
\begin{vmatrix}
z_1^0 & \cdots & z_{j-1}^0 & z_{j+1}^0 & \cdots & z_n^0 \\
\cdots & & & & & \\
z_1^{i-2} & \cdots & z_{j-1}^{i-2} & z_{j+1}^{i-2} & \cdots & z_n^{i-2} \\
z_1^i & \cdots & z_{j-1}^i & z_{j+1}^i & \cdots & z_n^i \\
\cdots & & & & & \\
z_1^{n-2} & \cdots & z_{j-1}^{n-1} & z_{j+1}^{n-1} & \cdots & z_n^{n-1}
\end{vmatrix},
$$

the cofactor of a_{ij}, is a homogeneous polynomial of degree $\dfrac{n(n-1)}{2}-(i-1)$ in $\hat{z}^j = (z_1, \cdots, z_{j-1}, z_{j+1}, \cdots, z_n)$. Moreover, A_{ij} is skew symmetric in its arguments, that is, for any $k_1 \neq k_2$,

$$A_{ij}(z_1, \cdots, z_{k_1}, \cdots, z_{k_2}, \cdots, z_n)$$
$$= -A_{ij}(z_1, \cdots, z_{k_2}, \cdots, z_{k_1}, \cdots, z_n).$$

By the basic property of the determinant, if $z_k = z_{k'}$ for $k \neq k'$ then $A_{ij} = 0$. Thus for $k \neq k'$, if $k \neq j$, $k' \neq j$ then $z_k - z_{k'}$ is a factor of A_{ij}. Thus we have

$$(\ast\ast\ast) \qquad A_{ij} = (-1)^{i+j} \prod_{\substack{k>k' \\ k\neq j, k'\neq j}} (z_k - z_{k'})Q(\hat{z}^j),$$

where Q is a homogeneous polynomial in $\hat{z}^j$, and

$$\begin{aligned}
\text{degree of } Q &= \text{degree of } A_{ij} - \text{degree of } \prod_{\substack{k>k' \\ k\neq j, k'\neq j}} (z_k - z_{k'}) \\
&= \left(\frac{n(n-1)}{2} - (i-1)\right) - \frac{(n-1)(n-2)}{2} \\
&= n - i.
\end{aligned}$$

It is clear that the highest degree of the terms of A_{ij} containing z_l for $l \neq j$ is $n-1$ and the highest degree of the terms of

$$\prod_{\substack{k>k' \\ k\neq j, k'\neq j}} (z_k - z_{k'})$$

containing z_l is $n-2$. Comparing the coefficients of z_l for $l \neq j$ on both sides of $(\ast\ast\ast)$, we see that the highest degree of the terms of $Q(\hat{z}^j)$ containing z_l is $n-2$. Moreover, since

$$\prod_{\substack{k>k' \\ k\neq j, k'\neq j}} (z_k - z_{k'})$$

is skew symmetric, $Q(\hat{z}^j)$ is a symmetric polynomial in its arguments, that is, for $k_1 \neq k_2$,

$$Q(z_1, \cdots, z_{k_1}, \cdots, z_{k_2}, \cdots, z_n)$$
$$= Q(z_1, \cdots, z_{k_2}, \cdots, z_{k_1}, \cdots, z_n).$$

Thus we have $Q(\hat{z}^j) = \alpha\sigma^j_{n-i}$, where α is a constant. Comparing the coefficients of $(***)$ again, we know that $\alpha = (-1)^{n-i}$.

In summary, we have

$$A_{ij} = (-1)^{n+j} \prod_{\substack{k>k' \\ k\neq j, k'\neq j}} (z_k - z_{k'})\sigma^j_{n-i}. \P$$

Let $B = (b_{ij}) = P'(z)$. Then the cofactor of b_{ij} is $B_{ij} = \dfrac{n!}{i}A_{ij}$ and $\det B = \det P'(z) = n! \prod_{k>k'}(z_k - z_{k'})$. Denote $B^{-1} = (\tilde{b}_{ij})$. Then

$$
\begin{aligned}
\hat{b}_{ij} &= \frac{n!}{j}\frac{A_{ij}}{\det B} \\[2mm]
&= \frac{n!}{j}\frac{(-1)^{n+i}\sigma^i_{n-j}\prod_{\substack{k>k' \\ k\neq i, k'\neq i}}(z_k - z_{k'})}{n! \prod_{k>k'}(z_k - z_{k'})} \\[2mm]
&= \frac{1}{j}\frac{(-1)^{n+i}}{(-1)^{i-1}}\frac{\sigma^i_{n-j}}{\prod_{k\neq i}(z_k - z_i)} \\[2mm]
&= (-1)^{n+1}\frac{1}{j}\frac{\sigma^i_{n-j}}{\prod_{k\neq i}(z_k - z_i)}.
\end{aligned}
$$

Thus the path Γ_{a_μ} is the unique solution of the following initial value problem:

$$
\begin{cases}
\dfrac{dz_j}{dt} = -\displaystyle\sum_{j=1}^{n}\tilde{b}_{ij}(P_j(a) - s_j) \\[4mm]
\quad = \dfrac{(-1)^n}{j}\dfrac{1}{\prod_{k\neq i}(z_k - z_i)}\displaystyle\sum_{j=1}^{n}\sigma^i_{n-j}(P_j(a) - s_j) \\[4mm]
z_i(0) = a_{\mu(i)},
\end{cases}
$$

where $a_\mu = (a_{\mu(1)}, \cdots, a_{\mu(i)}, \cdots, a_{\mu(n)})$.

Remark 3.10   Let $P : \mathbf{C}^n \to \mathbf{C}^n$ be a polynomial mapping and $q_j > 0$ be the degree of its j-th component P_j for $j = 1, 2, \cdots, n$. $q = \prod_{j=1}^n q_j$ is called the total degree of P. Let $\overline{P}$ be the homogeneous polynomial mapping of P with highest degree. If $\overline{P}$ has nontrivial zeros then P is called a deficient polynomial mapping. In other words, P is deficient if the number of the isolated zeros of P is less than its total degree q. For a deficient polynomial mapping P, if we use the homotopies as shown in Section 2 to locate the all isolated zeros of P, some of or very often even majority of the homotopy curves will diverge to infinity. This causes the proliferation of computations. Recently, many works appeared devoted to locating efficiently the all isolated zeros of deficient polynomial mappings and have obtained many significant results. Refer to [Li, Sauer & Yorke, 1987a, 1987b], [Li & Wang, 1991] for detail. Besides, there are some results on zero distributions, see [Wang, 1989], [Gao & Wang, 1991] and [Gao & Wang, 1992].

Chapter 8

Piecewise Linear Algorithms

Let $H : \mathbf{R}^m \times [0,1] \to \mathbf{R}^m$ be a differentiable mapping and $\mathbf{T}^\delta$ be a simplicial triangulation of $\mathbf{R}^m \times [0,1]$ with the mesh size δ (cf. [Todd, 1976] or [Wang, 1986]). And also let $\Phi_\delta : \mathbf{R}^m \times [0,1] \to \mathbf{R}^m$ be the piecewise linear approximation of H with respect to $\mathbf{T}^\delta$, that is, Φ_δ coincides with H at the vertices of $\mathbf{T}^\delta$ and Φ_δ is affine in every simplex of $\mathbf{T}^\delta$.

Regard $\mathbf{C}^n = \mathbf{R}^{2n}$ as in Chapter 7. Let $P : \mathbf{C}^n \to \mathbf{C}^n$ be a polynomial mapping. Define the linear homotopy $H(z,t) = tP(z) + (1-t)Q(z)$, where $Q_j(z) = z_j^{q_j} - b_j^{q_j}$, here q_j is the degree of the j-th component P_j of P and $b_j \neq 0$, $j = 1, \cdots, n$. Then we have shown in Chapter 7 that for almost every polynomial mapping P, $H^{-1}(0)$ consists of $q = \prod_{j=1}^n q_j$ distinct differentiable paths.

Let $\{\mathbf{T}^{\delta_i} : i = 1, 2, \cdots\}$ be a sequence of regular simplicial triangulations of $\mathbf{R}^{2n} \times [0,1]$ with $\delta_i \to 0$ as $i \to \infty$ (cf. §8.2). In this chapter, we will prove that for almost every polynomial mapping P, $\Phi_{\delta_i}^{-1}(0)$ consists of only piecewise linear paths and every path doesn't intersect with all the v-dimensional skeletons of $\mathbf{T}^{\delta_i}$, $0 \leq v \leq 2n - 1$. When δ_i is small enough, the piecewise linear paths of $\Phi_{\delta_i}^{-1}$ approximate the differentiable paths of $H^{-1}(0)$, and thus the zeros of $\Phi_{\delta_i}(\cdot, 1)$ can be regarded as the numerical zeros of $H(\cdot, 1) = P(\cdot)$.

To locate a piecewise linear path of $\Phi_{\delta_i}^{-1}(0)$, the simplicial homotopy algorithm utilizes the complementary pivoting procedure to produce a sequence of simplices

$$\sigma_0^{2n}, \sigma_0^{2n+1}, \sigma_1^{2n}, \sigma_1^{2n+1}, \cdots, \sigma_{l-1}^{2n+1}, \sigma_l^{2n}$$

which contains the piecewise linear path, where $\sigma_0^{2n} \subset \sigma_0^{2n+1}$, $\sigma_j^{2n} = \sigma_{j-1}^{2n+1} \cap \sigma_j^{2n+1}$, $j = 1, \cdots, l-1$, $\sigma_l^{2n} \subset \sigma_{l-1}^{2n+1}$, and $\sigma_0^{2n} \subset \mathbf{C}^n \times \{0\}$, $\sigma_l^{2n} \subset \mathbf{C}^n \times \{1\}$. Let $(u^j, t_j) \in \sigma_j^{2n}$ be the end of the line segment of the path in σ_j^{2n+1}. Then $\Phi_{\delta_i}(u^j, t_j) = 0$, $j = 0, 1, \cdots, l$ and (u^l, t_l) is a numerical zero of $P(\cdot)$. This is the main idea of simplicial homotopy algorithms.

§1.   Zeros of PL Mapping and Their Indexes

In the first two sections, we follow [Eaves, 1976] and [Eaves & Scarf, 1976].

Definition 1.1   Let $X_1, \cdots, X_k$ be the usual $(m + 1)$-dimensional closed polyhedral convex sets in $\mathbf{R}^{n+1}$ with the intersection of every pair of the sets containing no their interior points. $X = \bigcup_{i=1}^k X_i$ is called a polyhedron and X_i is called a piece of X.

Example 1.2   As in Fig. 8.1, the polyhedron X consists of convex pieces. But X itself is not convex. Besides, the intersection of some pair of pieces may not be a face of the pieces.

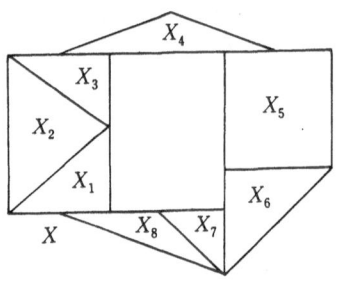

Figure 8.1

Fig. 8.1 shows also that X may not be simple connected. It is easy to design an example of a polyhedron which is even not connected.

Definition 1.3   Let $X_1, \cdots, X_k, X$ be as in Definition 1.1. If the mapping $F : X \to \mathbf{R}^m$ is continuous and is linear in every piece X_i, that is, for all $x, x' \in X_i$,

$$F(\alpha x + (1 - \alpha)x') = \alpha F(x) + (1 - \alpha)F(x'),$$

$i = 1, \cdots, k$. Then F is called a piecewise linear mapping on X.

Lemma 1.4   *If the restriction of the piecewise linear mapping F on*

X_i has full rank m then

$$F^{-1}(c) \cap X_i$$

is either empty or a line segment (may degenerate into a point) with two ends lying in two m-dimensional faces of X_i, where $F^{-1}(c) = \{x \in X : F(x) = c\}$.

Proof. Since the rank of the restriction of F on X_i is m, F can be extended to a linear mapping $\tilde{F}$ from $\mathbf{R}^{m+1}$ to $\mathbf{R}^m$. It is clear that $\tilde{F}^{-1}(c)$, the solution set of $\tilde{F} = c$, is a line. Thus, $F^{-1}(c) \cap X_i = \tilde{F}^{-1}(c) \cap X_i$ is either an empty set or a line segment with its ends in two m-dimensional faces of X_i(the segment may degenerate into one point). ¶

Remark 1.5   In Lemma 1.4, if the restriction of F on X_i has rank r which is less than m then $F^{-1}(c) \cap X_i$ is the intersection of X_i with some $((m+1) - r)$-dimensional hyperplane.

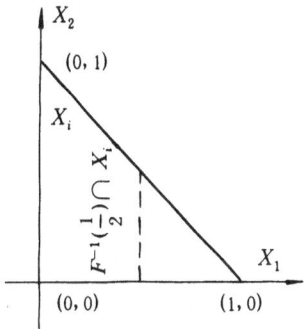

Figure 8.2

Example 1.6   As in Fig. 8.2, let X_i be the triangle in $\mathbf{R}^2$ with vertices $(0,0)$, $(0,1)$ and $(1,0)$. Define the linear mapping $F : X_i \to \mathbf{R}$ by $F(x_1, x_2) = x_1$. It is clear that $\text{rank} F = \text{rank}(1,0) = 1$ and

$$
\begin{aligned}
&F^{-1}(c) \cap X_i \\
={} &\{(x_1, x_2) \in X : F(x_1, x_2) = x_1 = c\} \cap X_i \\
={} &\{(c, x_2) : x_2 \in \mathbf{R}\} \cap X_i \\
={} &\begin{cases}
\emptyset, & \text{if } c \in (-\infty, 0) \cup (1, \infty), \\
\text{a line segment in the one-dimensional face} \\
\quad \langle (0,0), (0,1) \rangle, \text{ if } c = 0, \\
\text{a line segment with two ends in two different} \\
\quad \text{one-dimensional faces,} \quad \text{if } c \in (0,1), \\
(1,0), \quad \text{if } c = 1.
\end{cases}
\end{aligned}
$$

As in differential topology, we introduce

Definition 1.7   Let $X_1, \cdots, X_k, X$ and $F : X \to \mathbf{R}^m$ be as in Definition 1.3. If there is a point x lying in a k-dimensional face of some piece X_i, $k \leq m - 1$, such that $F(x) = c$ then $c \in \mathbf{R}^m$ is called a critical value of F. If c is not a critical value of F then c is called a regular value of F.

It is clear that if $c \notin F(X)$ then c is automatically a regular value of F.

Example 1.8   Let X consist of the four triangles in $\mathbf{R}^2$ as shown in Fig. 8.3. Define $F : X \to \mathbf{R}$ by $F(x_1, x_2) = x_1 + x_2$.

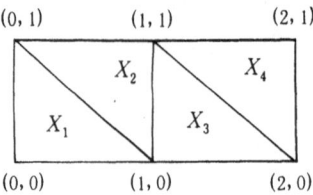

Figure 8.3

Since

$$F(0, 1) = 1,$$

$$F(1, 1) = 2,$$

$$F(2, 1) = 3,$$

$$F(0, 0) = 0,$$

$$F(1, 0) = 1,$$

$$F(2, 0) = 2.$$

It is clear that the critical values of F are $\{0, 1, 2, 3\}$ and its regular values are $\mathbf{R} - \{0, 1, 2, 3\}$. Furthermore,

$$F^{-1}(0) = \{(0, 0)\},$$

$$F^{-1}(1) = \{\alpha(0, 1) + (1 - \alpha)(1, 0) : 0 \leq \alpha \leq 1\},$$

$$F^{-1}(2) = \{\alpha(1, 1) + (1 - \alpha)(2, 0) : 0 \leq \alpha \leq 1\},$$

$$F^{-1}(3) = \{(2, 1)\}.$$

Lemma 1.9   Let $X_1, \cdots, X_k, X$ and $F : X \to \mathbf{R}^m$ be as in Definition 1.3. Let also c be a regular value of F. If $F^{-1}(c) \bigcap X_i \neq \emptyset$ then

$F^{-1}(c) \cap X_i$ *is a line segment with two ends lying respectively in interiors of two different m-dimensional faces of* X_i.

Proof. Let $F = Ax + a$ for every $x \in X_i$, where A is an $m \times (m+1)$ matrix. If rank$A = r < m$ then $\{x \in \mathbf{R}^{m+1} : Ax + a = c\}$ is a subspace of $\mathbf{R}^{m+1}$ with dimension $(m+1) - r \geq 2$. Since $F^{-1}(c) \cap X_i \neq \emptyset$, the intersection of $F^{-1}(c)$ with some $(m-1)$-dimensional face of X_i is not empty. This contradicts the regularity of c. Thus, rank$A = m$. Since rank$A = m$, by Lemma 1.4, the solution set of $Ax + b = c$ in X_i is a line segment. This segment can't be contained in an m-dimensional face of X_i since, otherwise, the segment must intersect with some $(m-1)$-dimensional face of X_i. Therefore, the two ends of the segment must lie in the interiors of two different m-dimensional faces of X_i. ¶

Theorem 1.10 *Let* $F : X \to \mathbf{R}^m$ *be a piecewise linear mapping. If* c *is a regular value of* F *then* $F^{-1}(c)$ *consists of a finite number of piecewise linear paths and loops which can't intersect each other. Furthermore, the intersection of every path with the boundary* ∂X *of* X *is two points lying in the interiors of two different m-dimensional faces of* X*, and every loop can't intersect with* ∂X*. (cf.Fig.8.4)*

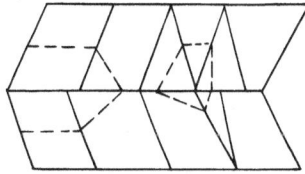   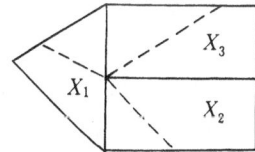

Figure 8.4   Figure 8.5

Proof. By Lemma 1.9, if $F^{-1}(c) \cap X_i \neq \emptyset$ then $F^{-1}(c) \cap X_i$ is a line segment with its two ends lying respectively in the interiors of two different m-dimensional faces of X_i. If one end is not in ∂X then it belongs to the interior of some m-dimensional face of other unique piece X_j adjacent to X_i. It is clear that $F^{-1}(c) \cap X_j \neq \emptyset$ and it is also a line segment. Continue this procedure, after finite number of steps, we obtain a piecewise linear path. If two ends of the path lie respectively in the interiors of two different m-dimensional faces then the ends must be in ∂X; If two ends lie in the interior of same m-dimensional face then the two ends must be same (otherwise, the line in $F^{-1}(c)$ linking the two ends must intersect with some $(m-1)$-dimensional face of X, this contradicts to the regularity of c). Now, it is easy to see that the end is an interior point of X. Hence the path becomes a loop which does not intersect with ∂X.

Notice that the above paths or loops are connected components of $F^{-1}(c)$. Since X consists of a finite number of pieces, the paths and loops are all finite, and the number of the paths and loops is also finite. ¶

Similarly to Sard's Theorem of differential topology, we have

Theorem 1.11   *Let $F : X \to \mathbf{R}^m$ be a piecewise linear mapping. Then the set of critical values of F is a closed set in $\mathbf{R}^m$ and is contained in the union of a finite number of $(m-1)$-dimensional hyperplanes.*

Proof.   Let Y_{i_j}, $j = 1, \cdots, l_i$ be the all $(m-1)$-dimensional faces of X_i. Since F is linear on X_i, $F(Y_{i_j}) \subset \pi_{i_j}$, where π_{i_j} is an $(m-1)$-dimensional hyperplane of $\mathbf{R}^m$. Thus, the set of the critical values of F satisfies

$$F\left(\bigcup_{i=1}^{k}\bigcup_{j=1}^{l_i} F(Y_{i_j})\right) = \bigcup_{i=1}^{k}\bigcup_{j=1}^{l_i} F(Y_{i_j}) \subset \bigcup_{i=1}^{k}\bigcup_{j=1}^{l_i} \pi_{i_j}.$$

It is clear that $F(Y_{i_j}) \subset \pi_{i_j}$ is a closed subset. Thus, the set of the critical values of F is closed. ¶

Theorem 1.12   *Let Y be an m-dimensional face of the piece X_i and x is an interior point of the face. If $\dim F(Y) = m$ then every neighborhood U of x in Y contains some point x' such that $c' = F(x')$ is a regular value of F.*

Proof.   It is clear that $\dim F(U) = m$ since $\dim F(Y) = m$. By Theorem 1.11, the set of critical values of F satisfies

$$F\left(\bigcup_{i=1}^{k}\bigcup_{j=1}^{l_i} Y_{i_j}\right) \subset \bigcup_{i=1}^{k}\bigcup_{j=1}^{l_i} \pi_{i_j},$$

where each π_{i_j} is an $(m-1)$-dimensional hyperplane of $\mathbf{R}^m$. Thus there is an $x' \in U$ such that

$$c' = F(x') \notin \bigcup_{i=1}^{k}\bigcup_{j=1}^{l_i} \pi_{i_j}.$$

This means that c' is a regular value of F. ¶

The result of Theorem 1.12 allows us to deal with the degenerate cases by small perturbations.

Definition 1.13   Let $\{b^1, \cdots, b^m\}$ and $\{\tilde{b}^1, \cdots, \tilde{b}^m\}$ be two bases of $\mathbf{R}^m$ satisfying

$$
\begin{pmatrix} b^1 \\ \vdots \\ b^m \end{pmatrix} = \begin{pmatrix} \lambda_{11} & \cdots & \lambda_{1m} \\ \cdots & \cdots & \cdots \\ \lambda_{m1} & \cdots & \lambda_{mm} \end{pmatrix} \begin{pmatrix} \tilde{b}^1 \\ \vdots \\ \tilde{b}^m \end{pmatrix}.
$$

If $\det(\lambda_{ij}) > 0$ then $\{b^1, \cdots, b^m\}$ and $\{\tilde{b}^1, \cdots, \tilde{b}^m\}$ are said to have the same orientation; Otherwise, $\{b^1, \cdots, b^m\}$ and $\{\tilde{b}^1, \cdots, \tilde{b}^m\}$ are said to have opposite orientations.

By Definition 1.13, we can classify the all bases of $\mathbf{R}^m$ into two classes. The bases in same class have same orientation. Sometimes, we assign the positive orientation to one class of the bases and the negative orientation to the other class.

Definition 1.14   Let $T : \mathbf{R}^m \to \tilde{\mathbf{R}}^m = \mathbf{R}^m$ be a linear transformation. Choose $\{b^1, \cdots, b^m\}$ and $\{\tilde{b}^1, \cdots, \tilde{b}^m\}$ to be respectively the bases of $\mathbf{R}^m$ and $\tilde{\mathbf{R}}^m$ with positive orientation. If T transforms the bases with positive orientation into the bases with positive orientation, then define the index of T to be 1; Otherwise, define the index of T to be -1.

Let

$$
\begin{pmatrix} Tb^1 \\ \vdots \\ Tb^m \end{pmatrix} = \begin{pmatrix} \lambda_{11} & \cdots & \lambda_{1m} \\ \cdots & \cdots & \cdots \\ \lambda_{m1} & \cdots & \lambda_{mm} \end{pmatrix} \begin{pmatrix} \tilde{b}^1 \\ \vdots \\ \tilde{b}^m \end{pmatrix}.
$$

It is clear that the index of T is equal to 1 if and only if $\det(\lambda_{ij}) > 0$, and the index of T is equal to -1 if and only if $\det(\lambda_{ij}) < 0$

It is noteworthy that the index of a linear transformation is dependent on the chosen positive orientation bases of $\mathbf{R}^m$ and $\tilde{\mathbf{R}}^m$. If we choose $\{-b^1, b^2, \cdots, b^m\}$ and $\{-\tilde{b}^1, \tilde{b}^2, \cdots, \tilde{b}^m\}$ as the bases of $\mathbf{R}^m$ and $\tilde{\mathbf{R}}^m$ with positive orientation, then the index of T with respect to the bases is same as in Definition 1.14. But if we choose $\{-b^1, b^2, \cdots, b^m\}$ and $\{\tilde{b}^1, \tilde{b}^2, \cdots, \tilde{b}^m\}$ or $\{b^1, b^2, \cdots, b^m\}$ and $\{-\tilde{b}^1, \tilde{b}^2, \cdots, \tilde{b}^m\}$ as the bases of $\mathbf{R}^m$ and $\tilde{\mathbf{R}}^m$ then the index of T with respect to the bases is opposite to the one in Definition 1.14.

Let $F : X \to \mathbf{R}^m$ be as in Theorem 1.10 and c be a regular value of F. Of course, we can assign every path or every loop in $F^{-1}(c)$ two different orientations as shown in Figs. 8.6 and 8.7.

Consider the piece X_i. If $F^{-1}(c) \cap X_i \neq \emptyset$ then it is a line segment with two ends lying respectively in two different m-dimensional faces of

X_i. This line segment is a part of some path or some loop oriented, and thus we can choose a vector $q \in \mathbf{R}^{m+1}$ (if necessary, set $|q| = 1$) to indicate its positive orientation and $-q$ to indicate its negative orientation(cf. Fig. 8.8).

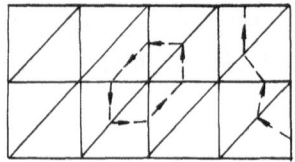

Figure 8.6

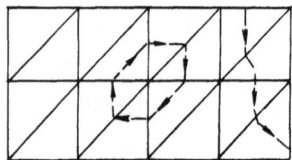

Figure 8.7

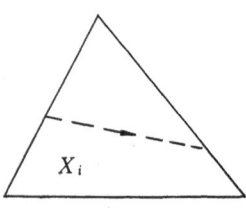

Figure 8.8

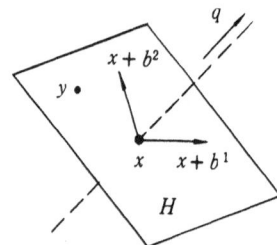

Figure 8.9

Let $H \subset \mathbf{R}^{m+1}$ be an m-dimensional hyperplane which intersects transversely with $F^{-1}(c) \cap X_i$ at x, and let the positive orientation of the line segment $F^{-1}(c) \cap X_i$ be indicated by $q \in \mathbf{R}^{m+1}$. Denote the subspace(cf. Fig. 8.9)

$$V_H = \{y - x : y \in H\}.$$

Let $\{b^1, \cdots, b^m\}$ be a base of V_H. If

$$\det \begin{pmatrix} b_1^1 & b_1^2 & \cdots & b_1^m & q_1 \\ b_2^1 & b_2^2 & \cdots & b_2^m & q_2 \\ \cdots & & & & \\ b_{m+1}^1 & b_{m+1}^2 & \cdots & b_{m+1}^m & q_{m+1} \end{pmatrix} > 0,$$

then $\{b^1, \cdots, b^m\}$ is called a base of V_H with positive orientation; Otherwise, if $\det(b^1, \cdots, b^m, q) < 0$ then $\{b^1, \cdots, b^m\}$ is called a base with negative orientation. In other words, the orientation of the base $\{b^1, \cdots, b^m\}$

of V_H is determined by the orientation of the base of $\mathbf{R}^{m+1}$ and the orientation q of the line segment running through H.

Let F be a piecewise linear mapping and c be a regular value of F. Assume $F(x) = Ax + a$ in X_i, where A is an $m \times (m+1)$ matrix. Then F induces a linear mapping $A : V_H \to \mathbf{R}^m$ defined by

$$A(y - x) = F(y) - F(x) \quad \text{for all } x, y \in H \text{ and } y - x \in V_H.$$

By Lemma 1.9, rank$A = m$. Thus

$$
\begin{aligned}
m &= \text{rank}(Ab^1, \cdots, Ab^m, Aq) \\
&= \text{rank}(Ab^1, \cdots, Ab^m, 0) \\
&= \text{rank}(Ab^1, \cdots, Ab^m),
\end{aligned}
$$

that is, $\{Ab^1, \cdots, Ab^m\}$ is a base of $\mathbf{R}^m$, where $Aq = 0$ because

$$
\begin{aligned}
0 &= c - c = F(x) - F(x + \lambda q) \\
&= (Ax + a) - (A(x + \lambda q) + a) \\
&= \lambda Aq, \text{ for } \lambda \neq 0 \text{ and } x + \lambda q \in X_i.
\end{aligned}
$$

If we choose the standard base of $\mathbf{R}^m$ as the base with positive orientation then, by Definition 1.14, the index $\mathcal{D}_A$ of the linear mapping $A : V_H \to \mathbf{R}^m$ can be calculated by

$$
\mathcal{D}_A = \begin{cases}
1, & \text{if } \det(b^1, \cdots, b^m, q) \cdot \det(Ab^1, \cdots, Ab^m) > 0, \\
-1, & \text{if } \det(b^1, \cdots, b^m, q) \cdot \det(Ab^1, \cdots, Ab^m) < 0.
\end{cases}
$$

Lemma 1.15 *The index of the linear mapping $A : V_H \to \mathbf{R}^m$ is same for every m-dimensional hyperplane H intersecting transversely with $F^{-1}(c) \cap X_i$ at x. Furthermore, the index is also independent of the choice of x on $F^{-1}(c) \cap X_i$.*

Proof. Notice that

$$
\det \begin{pmatrix}
a_{11} & a_{12} & \cdots & a_{1,m+1} \\
\cdots & & & \\
a_{m1} & a_{m2} & \cdots & a_{m,m+1} \\
q_1 & q_2 & \cdots & q_{m+1}
\end{pmatrix}
\begin{pmatrix}
b_1^1 & b_1^2 & \cdots & b_1^m & q_1 \\
b_2^1 & b_2^2 & \cdots & b_2^m & q_2 \\
\cdots & & & & \\
b_{m+1}^1 & b_{m+1}^2 & \cdots & b_{m+1}^m & q_{m+1}
\end{pmatrix}
$$

$$
= \det \begin{pmatrix}
 & & & & 0 \\
Ab^1 & Ab^2 & \cdots & Ab^m & \vdots \\
 & & & & 0 \\
q \cdot b^1 & q \cdot b^2 & \cdots & q \cdot b^m & |q|^2
\end{pmatrix}
$$

$$
= |q|^2 \cdot \det(Ab^1, \cdots, Ab^m).
$$

We know that $\det(b^1, \cdots, b^m, q) \cdot \det(Ab^1, \cdots, Ab^m)$ and

$$
\det \begin{pmatrix} a_{11} & a_{12} & \cdots & a_{1,m+1} \\ \cdots & & & \\ a_{m1} & a_{m2} & \cdots & a_{m,m+1} \\ q_1 & q_2 & \cdots & q_{m+1} \end{pmatrix} = \det \begin{pmatrix} A \\ q \end{pmatrix}
$$

have same sign. This implies that

$$
\mathcal{D}_A = \begin{cases} 1, & \text{if } \det \begin{pmatrix} A \\ q \end{pmatrix} > 0, \\[2ex] -1, & \text{if } \det \begin{pmatrix} A \\ q \end{pmatrix} < 0. \end{cases}
$$

It is clear that $\mathcal{D}_A$ is independent of H and the choice of x.¶

Theorem 1.16   *If c is a regular value of the piecewise linear mapping F then F keeps the same index at different points on the same path or loop of $F^{-1}(c)$*

Proof.   Let x^* lie in the interior of the intersection of two adjacent pieces X_1 and X_2. By Lemma 1.15, we need only to prove that the indexes at x^* with respect to X_1 and with respect to X_2 are same. Let $F(x) = A^i x + a^i$ for $x \in X_i$ and q^i be the positive direction of the path in X^i(cf. Fig. 8.10), $i = 1, 2$. Also let $\{b^1, \cdots, b^m\}$ be a base of V_H.

Since q^1 and q^2 point to the same side of H we have

$$
q^2 = \sum_{j=1}^m \alpha_j b^j + \theta q^1, \quad \theta > 0.
$$

Thus by the basic properties of determinants,

$$
\begin{aligned}
\det(b^1, \cdots, b^m, q^2) &= \det\left(b^1, \cdots, b^m, \sum_{j=1}^m \alpha_j b^j + \theta q^1\right) \\
&= \theta \cdot \det(b^1, \cdots, b^m, q^1).
\end{aligned}
$$

Notice that for $x \in H$

$$
A^1 x + a^1 = A^2 x + a^2
$$

and

$$
A^1(x^* + b^j) + a^1 = A^2(x^* + b^j) + a^2,
$$

thus we have

$$A^1 b^j = A^2 b^j, \text{ for } j = 1, \cdots, m$$

and

$$\det(A^1 b^1, \cdots, A^1 b^m) = \det(A^2 b^1, \cdots, A^2 b^m).$$

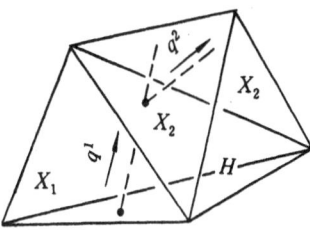

Figure 8.10

In summary, we obtain that

$$\det(b^1, \cdots, b^m, q^2) \cdot \det(A^1 b^1, \cdots, A^1 b^m)$$
$$= \theta \cdot \det(b^1, \cdots, b^m, q^1) \cdot \det(A^2 b^1, \cdots, A^2 b^m)$$

and

$$\det(b^1, \cdots, b^m, q^1) \cdot \det(A^2 b^1, \cdots, A^2 b^m)$$

have same sign. This completes the proof. ¶

Now know that the index discussed in Definition 1.14, Lemma 1.15 and Theorem 1.16 depends only on the direction of the path. We call it the index of oriented curves. It is clear that the sign of the index changes as the direction of the path changes. Next we will give the definition of the boundary index.

Definition 1.17 Let x^* be a boundary point of $F^{-1}(c)$. Suppose that x^* lies in the interior of some m-dimensional face H of some piece X_i. Let $F(x) = Ax + a$ for $x \in X_i$ and $\{b^1, \cdots, b^m\}$ be a base of V_H. Let q be any fixed direction at x^* pointing into X_i (q may be different from the direction of the path). Define the boundary index $\mathcal{D}(x^*)$ to be

$$\mathcal{D}(x^*) = \begin{cases} 1, & \text{if } \det(b^1, \cdots, b^m, q) \cdot \det(A b^1, \cdots, A b^m) > 0, \\ -1, & \text{if } \det(b^1, \cdots, b^m, q) \cdot \det(A b^1, \cdots, A b^m) < 0. \end{cases}$$

It is noteworthy that, similar to the proof of Theorem 1.16, we can prove that the sign $\det(b^1, \cdots, b^m, q) \cdot \det(Ab^1, \cdots, Ab^m)$ is independent of the choice of the direction q at x^*.

Theorem 1.18 (Index Theorem)   *Let c be a regular value of the piecewise linear mapping $F : X \to \mathbf{R}^m$. Then*

$$\sum_{x^* \in F^{-1}(c) \cap \partial X} \mathcal{D}(x^*) = 0.$$

Proof. Let x^1 and x^2 be two ends of some path in $F^{-1}(c)$. Then $x^1, x^2 \in F^{-1}(c) \cap \partial X$. Also let q^1 and q^2 be respectively the directions at x^1 and x^2 coinciding with the direction of the path (cf. Fig. 8.11).

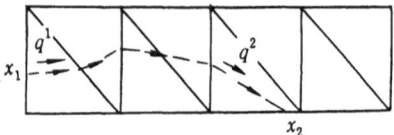

Figure 8.11

By Theorem 1.16, we have

$$\mathcal{D}(x^1) = \mathcal{D}_A(x^1) = \mathcal{D}_A(x^2) = -\mathcal{D}(x^2),$$

that is,

$$\mathcal{D}(x^1) + \mathcal{D}(x^2) = 0.$$

This implies that

$$\sum_{x^* \in F^{-1}(c) \cap \partial X} \mathcal{D}(x^*) = 0. \P$$

§2.   PL Approximations

We need the general-case concept of simplicial triangulations.

It is well-known that a collection G of m-simplices is a simplicial triangulation (or a triangulation in short) of some convex subset C of the Euclidean space if

(1) C is the union of all simplices in G;

and

(2) the intersection of two simplices in G is either empty or a common face.

Notice that $\{X_1, \cdots, X_k\}$ in Definition 1.1 is not a simplicial triangulation.

Let $\mathbf{T}$ be a triangulation of $\mathbf{R}^m \times [0, 1]$ and $\Phi : \mathbf{R}^m \times [0, 1] \to \mathbf{R}^m$ be a piecewise linear mapping with respect to $\mathbf{T}$. If c is a regular value of Φ, from arguments similar to Lemma 1.9 and Theorem 1.10, we know that $\Phi^{-1}(c)$ consists of a countable number of paths and loops, and every path or loop is made of a countable number of line segments. Furthermore, the ends of the paths must lie in

$$\partial(\mathbf{R}^m \times [0, 1]) = (\mathbf{R}^m \times \{0\}) \bigcup (\mathbf{R}^m \times \{1\})$$

and every loop does not intersect with $\partial(\mathbf{R}^m \times [0, 1])$ (cf. Fig. 8.12).

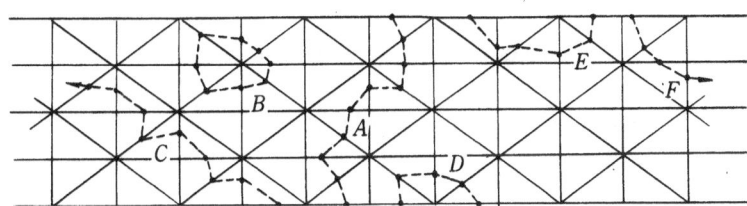

Figure 8.12

The most interesting is the path A in Fig. 8.12. It consists of a finite number of line segments, and every line segment belongs to a unique $(m+1)$-dimensional simplex σ_j^{m+1} with two ends respectively lying in the interiors of two different m-dimensional faces σ_j^m and σ_{j+1}^m of σ_j^{m+1}. Thus, the path A is uniquely determined by a finite number of simplices

$$\sigma_0^m, \sigma_0^{m+1}, \sigma_1^m, \sigma_1^{m+1}, \cdots, \sigma_{l-1}^m, \sigma_l^{m+1},$$

where $\sigma_0^m \subset \sigma_0^{m+1}$, $\sigma_j^m = \sigma_{j-1}^{m+1} \bigcap \sigma_j^{m+1}$, $j = 1, \cdots, l-1$, $\sigma_l^m \subset \sigma_{l-1}^{m+1}$, $\sigma_0^m \subset \mathbf{R}^m \times \{0\}$ and $\sigma_l^m \subset \mathbf{R}^m \times \{1\}$. Let $(u^j, t_j) \in \sigma_j^m$ be the end of the corresponding line segment. Then, it is clear that $\Phi(u^j, t_j) = 0$ for $j = 0, 1, \cdots, l$.

It is noteworthy that, though the path A does exist, it is rather difficult to locate the path due to the complicated nature of the mapping Φ. A feasible way is to locate the finite number of simplices by complementary pivoting algorithms, and thus determine the path A and a solution y^l of $\Phi(u,1) = 0$. If the solution $(u^0, t_0) = (u^0, 0) \in \sigma_0^m$ of $\Phi(u^0, 0) = 0$ is given, by the triangulation, there is a unique $(m+1)$-dimensional simplex σ_0^{m+1} with σ_0^m as its face. Notice that only one new vertex is needed to determine σ_0^{m+1} from σ_0^m. Since Φ is affine on σ_0^{m+1}, its zero set is a line segment with two ends $(u^0, 0)$ and (u^1, t_1). Let σ_1^m is the face of σ_0^{m+1} containing the end (u^1, t_1). Again by the triangulation, there is another unique $(m+1)$-dimensional simplex σ_1^{m+1} with σ_1^m as its face, then again, one can obtain the line segment, the zero set of Φ in σ_1^{m+1}. After finite number of steps, the path A can be entirely determined.

Theorem 2.1 *Let* $\mathbf{T}$ *be a triangulation of* $\mathbf{R}^m \times [0,1]$. *Let further* $\Phi : \mathbf{R}^m \times [0,1] \to \mathbf{R}^m$ *be the piecewise linear mapping with respect to* $\mathbf{T}$ *and* c *be a regular value of* Φ. *If a finite sequence of simplices*

$$\sigma_0^m, \sigma_0^{m+1}, \sigma_1^m, \sigma_1^{m+1}, \cdots, \sigma_{l-1}^m, \sigma_l^{m+1}$$

is given corresponding to a bounded piecewise linear path as above, then

(1)   *If both* σ_0^m *and* σ_l^m *lie in* $\mathbf{R}^m \times \{0\}$(*or* $\mathbf{R}^m \times \{1\}$), *then*

$$\operatorname{sgn} \det \Phi'_u|_{\sigma_0^m} = -\operatorname{sgn} \det \Phi'_u|_{\sigma_l^m};$$

(2)   *If* σ_0^m *and* σ_l^m *lie respectively in* $\mathbf{R}^m \times \{0\}$ *and* $\mathbf{R}^m \times \{1\}$ (*or* $\mathbf{R}^m \times \{1\}$ *and* $\mathbf{R}^m \times \{0\}$), *then*

$$\operatorname{sgn} \det \Phi'_u|_{\sigma_0^m} = \operatorname{sgn} \det \Phi'_u|_{\sigma_l^m}.$$

Proof.   (1)   Let

$$\Phi(u,t) = A^1 \begin{pmatrix} u \\ t \end{pmatrix} + a^1 \quad \text{for all } (u,t) \in \sigma_0^{m+1}$$

and

$$\Phi(u,t) = A^2 \begin{pmatrix} u \\ t \end{pmatrix} + a^2 \quad \text{for all } (u,t) \in \sigma_{l-1}^{m+1}.$$

Let q^1 and q^2 be the directions of the line segments of the path of $\Phi^{-1}(c)$ in σ_0^{m+1} and σ_{l-1}^{m+1}. Then by Lemma 1.15 and Theorem 1.16,

$$\det \Phi'_u|_{\sigma_0^m} = \det \begin{pmatrix} a_{11}^1 & \cdots & a_{1m}^1 \\ \cdots & & \\ a_{m1}^1 & \cdots & a_{mm}^1 \end{pmatrix}$$

$$= \det \begin{pmatrix} a_{11}^1 & \cdots & a_{1m}^1 & a_{1,m+1}^1 \\ \cdots & & & \\ a_{m1}^1 & \cdots & a_{mm}^1 & a_{m,m+1}^1 \\ 0 & \cdots & 0 & 1 \end{pmatrix}$$

$$= \det \begin{pmatrix} A^1 \\ q^1 \end{pmatrix}$$

$$= \det \begin{pmatrix} A^2 \\ q^2 \end{pmatrix} .$$

$$= -\det \begin{pmatrix} a_{11}^2 & \cdots & a_{1m}^2 & a_{1,m+1}^2 \\ \cdots & & & \\ a_{m1}^2 & \cdots & a_{mm}^2 & a_{m,m+1}^2 \\ 0 & \cdots & 0 & 1 \end{pmatrix}$$

$$= -\det \begin{pmatrix} a_{11}^2 & \cdots & a_{1m}^2 \\ \cdots & & \\ a_{m1}^2 & \cdots & a_{mm}^2 \end{pmatrix}$$

$$= -\det \Phi_u'|_{\sigma_l^m},$$

that is,

$$\text{sgn} \det \Phi_u'|_{\sigma_0^m} = -\text{sgn} \det \Phi_u'|_{\sigma_l^m}.$$

(2) The proof is similar to that of (1) and thus is omitted. ¶

Any pair of vertices of a simplex determines an edge of the simplex. An edge of a simplex in a triangulation is also called an edge of the triangulation.

Definition 2.2 Let $\mathbf{T}$ be a triangulation of $\mathbf{R}^m \times [0,1]$. The supreme length of the edges of the triangulation is called the mesh of the triangulation.

Let $\{\mathbf{T}^{\delta_i}\}$ be a sequence of triangulations of $\mathbf{R}^m \times [0,1]$ and δ_i be the mesh of $\mathbf{T}^{\delta_i}$ with $\delta_i \to 0$ as $i \to \infty$. See [Todd, 1976] or [Wang, 1986] for the existence of such triangulation. If for some fixed $\alpha > 0$ and for every $(m+1)$-dimensional simplex $\sigma = \{x^j\}_0^{m+1}$ in $\mathbf{T}^{\delta_i}(i = 1, 2, \cdots)$,

$$\det \begin{pmatrix} x^0 & \cdots & x^{m+1} \\ 1 & \cdots & 1 \end{pmatrix} = \det(x^1 - x^0, \cdots, x^{m+1} - x^0) \geq \alpha \delta_i^{m+1},$$

that is, the volume of the hyper-rectangle

$$\left\{ x \in \mathbf{R}^{m+1} : x = x^0 + \sum_{i=1}^{m+1} \lambda_i(x^i - x^0), 0 \leq \lambda_i \leq 1 \right\}$$

is greater than $\alpha \delta_i^{m+1}$, then $\{\mathbf{T}^{\delta_i}\}$ is called a sequence of regular triangulations, or a regular sequence of triangulations.

Definition 2.3   Let $H : \mathbf{R}^m \times [0,1] \to \mathbf{R}^m$ be a differentiable mapping and $\mathbf{T}^\delta$ be a triangulation of $\mathbf{R}^m \times [0,1]$ with mesh size δ. Define the piecewise linear approximation Φ_δ of H with respect to $\mathbf{T}^\delta$ as follows: If (u,t) is a vertex of $\mathbf{T}^\delta$ then $\Phi_\delta(u,t) = H(u,t)$; If

$$(u,t) = \sum_{j=0}^{m+1} \lambda_j(u^j, t^j) \in \sigma^{m+1} = \{(u^j, t_j)\}_0^{m+1} \in \mathbf{T}^\delta$$

with $\sum_{j=0}^{m+1} \lambda_j = 1$ and $\lambda_j \geq 0$ then

$$\Phi_\delta(u,t) = \sum_{j=0}^{m+1} \lambda_j \Phi_\delta(u^j, t_j) = \sum_{j=0}^{m+1} \lambda_j H(u^j, t_j).$$

It is easy to see that for $t = 0, 1$, the triangulation $\mathbf{T}^\delta$ induces a triangulation of $\mathbf{R}^m \times \{t\}$, and thus Φ_δ induces a piecewise linear approximation $\Phi_\delta(\cdot, t)$ of $H(\cdot, t)$ with respect to the induced triangulation.

Remark 2.4   It is clear that for any fixed $(u,t) \in \mathbf{R}^m \times [0,1]$, $\lim_{i \to \infty} \Phi_{\delta_i}(u,t) = H(u,t)$, where $\delta_i \to 0$ as $i \to \infty$.

Next several theorems will show that Φ_δ is indeed a good approximation of H.

Theorem 2.5   *Let $H : \mathbf{R}^m \times [0,1] \to \mathbf{R}^m$ be a differentiable mapping and $H'_u(\overline{u}, \overline{t})$ be nonsingular for a given $(\overline{u}, \overline{t}) \in H^{-1}(0)$. Let further $\{\mathbf{T}^{\delta_i}\}$ be a sequence of regular triangulations of $\mathbf{R}^m \times [0,1]$. Then there are an open neighborhood $w_{\overline{u}}$ of $\overline{u}$ and an $\epsilon > 0$ such that for $u \in w_{\overline{u}}$ and $0 < \delta_i \leq \epsilon$,*

$$\operatorname{sgn} \det (\Phi_{\delta_i})'_u(u, \overline{t}) = \operatorname{sgn} \det H'_u(\overline{u}, \overline{t}).$$

Proof.   Let $u \in \mathbf{R}^m$ and the simplex $\sigma = \{(u^j, t_j)\}_0^{m+1}$ contain $(u, \overline{t})$. Then

$$(u, \overline{t}) = \sum_{j=0}^{m+1} \lambda_j(u^j, t_j)$$

$$= (u^0, t_0) + \sum_{j=1}^{m+1} \lambda_j((u^j, t_j) - (u^0, t_0))$$

with $\sum_{j=0}^{m+1} \lambda_j = 1$ and $\lambda_j \geq 0$, and

$$\lambda = (\lambda_j) = \langle (u^j, t_j) - (u^0, t_0)\rangle^{-1}((u, \overline{t}) - (u^0, t_0)),$$

where $\langle (u^j, t_j) - (u^0, t_0) \rangle$ denote the $(m+1) \times (m+1)$ matrix with its j-th column $(u^j, t_j) - (u^0, T_0)$, $j = 1, \cdots, m+1$.

By Taylor formula, for $j = 1, 2, \cdots, m+1$,

$$\begin{aligned} \Phi_{\delta_i}(u^j, t_j) &= H(u^j, t_j) \\ &= H(u^0, t_0) + H'(u^0, \bar{t})((u^j, t_j) - (u^0, t_0)) + R^j, \end{aligned}$$

where

$$|R^j|_\infty \leq K_j |(u^j, t_j) - (u^0, \bar{t})|_\infty^2$$

with K_j a constant, and the norm $|\cdot|_\infty$ of $\mathbf{R}^{m+1}$ is defined by

$$|x|_\infty = \max_{1 \leq l \leq m+1} \{|x_l|\}.$$

Thus

$$\Phi_{\delta_i}(u^j, t_j) - \Phi_{\delta_i}(u^0, t_0) = H'(u^0, \bar{t})((u^j, t_j) - (u^0, t_0)) + R^j - R^0$$

and

$$\begin{aligned} \Phi_{\delta_i}(u, \bar{t}) &= \Phi_{\delta_i}(u^0, t_0) + \sum_{j=1}^{m+1} \lambda_j (\Phi_{\delta_i}(u^j, t_j) - \Phi_{\delta_i}(u^0, t_0)) \\ &= \Phi_{\delta_i}(u^0, t_0) + [H'(u^0, \bar{t}) \langle (u^j - t_j) - (u^0, t_0) \rangle \\ &\quad + \langle R^j - R^0 \rangle] \langle (u^j, t_j) - (u^0, t_0) \rangle^{-1} ((u, \bar{t}) - (u^0, t_0)) \\ &= \Phi_{\delta_i}(u^0, t_0) + [H'(u^0, \bar{t}) \\ &\quad + \langle R^j - R^0 \rangle \langle (u^j, t_j) - (u^0, t_0) \rangle^{-1}] ((u, \bar{t}) - (u^0, t_0)). \end{aligned}$$

It is now easy to obtain the derivatives of Φ_{δ_i} with respect to u as follows:

$$(\Phi_{\delta_i})'_u(u, \bar{t}) = H'_u(u^0, \bar{t}) + (\langle R^j - R^0 \rangle \langle (u^j, t_j) - (u^0, t_0) \rangle^{-1})_{m \times m},$$

where $A_{m \times m}$ denotes an $m \times m$ matrix obtained from the $m \times (m+1)$ matrix A deleted the last column. To prove

$$\lim_{i \to \infty} [(\Phi_{\delta_i})'_u(u, \bar{t}) - H'_u(u^0, t_0)] = 0,$$

we need only to prove that

$$\lim_{i \to \infty} \langle R^j - R^0 \rangle \langle (u^j, t_j) - (u^0, t_0) \rangle^{-1} = 0.$$

Since the sequence $\{\mathbf{T}^{\delta_i}\}$ is regular, for all δ_i, there is a positive number α such that

$$\det\langle(u^j, t_j) - (u^0, t_0)\rangle \geq \alpha\delta_i^{m+1}.$$

Let A_{ab} be the cofactor of the (a, b)-element of the matrix

$$\langle(u^j, t_j) - (u^0, t_0)\rangle.$$

Then its absolute

$$
\begin{aligned}
|A_{ab}| & = \left| \sum (-1)^p \prod_{\substack{j \neq b \\ k_j \neq a}} [(u^j, t_j)_{k_j} - (u^0, t_0)_{k_j}] \right| \\
& \leq \sum \prod_{\substack{j \neq b \\ k_j \neq a}} |(u^j, t_j) - (u^0, t_0)|_\infty \\
& \leq m!\delta_i^m \\
& = \frac{m!}{\alpha\delta_i}(\alpha\delta_i^{m+1}) \\
& \leq \frac{m!}{\alpha\delta_i} \det\langle(u^j, t_j) - (u^0, t_0)\rangle,
\end{aligned}
$$

where the sum $\sum$ runs through the all $m!$ permutations, and $p = 0$ if the permutation is even and $p = 1$ otherwise. It leads to the following estimate of the (a, b)-element of the matrix:

$$\frac{|A_{ab}|}{\det\langle(u^j, t_j) - (u^0 - t_0)\rangle} \leq \frac{m!}{\alpha\delta_i}.$$

Let $K = \max_{1 \leq j \leq (m+1)} K_j$. Then

$$
\begin{aligned}
|R^j - R^0|_\infty & \leq |R^j|_\infty + |R^0|_\infty \\
& \leq K_j |(u^j, t_j) - (u^0, \bar{t})|_\infty^2 + K_0 |(u^0, t_0) - (u^0, \bar{t})|_\infty^2 \\
& \leq 2K\delta_i^2.
\end{aligned}
$$

Thus, the value of the (a, b)-element of the matrix

$$\langle R^j - R^0 \rangle \langle(u^j, t_j) - (u^0, t_0)\rangle^{-1}$$

is

$$\left| \sum_h (R^h - R^0)_a \frac{A_{ab}}{\det\langle(u^j, t_j) - (u^0, t_0)\rangle} \right|$$

$$\leq \sum_h |R^h - R^0|_\infty \frac{m!}{\alpha \delta_i}$$

$$\leq 2(m+1)K\delta_i^2 \frac{m!}{\alpha \delta_i}$$

$$= \frac{2(m+1)(m!)K\delta_i}{\alpha}.$$

This implies that

$$\lim_{i \to \infty} \langle R^j - R^0 \rangle \langle (u^j, t_j) - (u^0, t_0) \rangle^{-1} = 0.$$

Let $U(\bar{u}, 2\delta)$ be a ball centered at $\bar{u}$ with radius 2δ. With an appropriate choice of $\delta > 0$, for every $v \in U(\bar{u}, 2\delta)$,

$$|H'_u(v, \bar{t}) - H'_u(\bar{u}, \bar{t})|_\infty < \frac{\eta}{2} \quad \text{for some } \eta > 0.$$

Let $w_{\bar{u}} = U(\bar{u}, \delta)$. When

$$0 < \delta_i \leq \min\left(\frac{\alpha\eta}{4(m+1)(m!)K}, \delta\right) = \epsilon$$

and $u^0 \in w_{\bar{u}}$, we have $u^0 \in U(\bar{u}, 2\delta)$. Thus,

$$|(\Phi_{\delta_i})'_u(u, \bar{t}) - H'_u(\bar{u}, \bar{t})|_\infty$$

$$\leq |(\Phi_{\delta_i})'_u(u, \bar{t}) - H'_u(u^0, \bar{t})|_\infty + |H'_u(u^0, \bar{t}) - H'_u(\bar{u}, \bar{t})|_\infty$$

$$\leq \frac{2(m+1)(m!)K\delta_i}{\alpha} + \frac{\eta}{2}$$

$$< \frac{\eta}{2} + \frac{\eta}{2} = \eta.$$

Since $H'_u(\bar{u}, \bar{t})$ is nonsingular for η small enough,

$$\text{sgn} \det(\Phi_{\delta_i})'_u(u, \bar{t}) = \text{sgn} \det H'_u(\bar{u}, \bar{t}). \P$$

Remark 2.6 The regularity of the sequence $\{\mathbf{T}^{\delta_i}\}$ guarantees that the volume of a simplex in $\mathbf{T}^{\delta_i}$ does not converge very fast to zero as δ_i approaches zero. Thus, $(\Phi_{\delta_i})'_u(u, \bar{t})$ can arbitrarily approximate $H'_u(\bar{u}, \bar{t})$.

Theorem 2.7 Let $H : \mathbf{R}^m \times [0, 1] \to \mathbf{R}^m$ be a differentiable mapping and $H'_u(u, t)$ be nonsingular at $(\bar{u}, \bar{t}) \in H^{-1}(0)$. Let further $\{\mathbf{T}^{\delta_i}\}$ be a sequence of regular triangulations of $\mathbf{R}^n \times [0, 1]$. If 0 is a regular value of

Φ_{δ_i} for $i = 1, 2, \cdots$, then there are an open neighborhood $w_{\bar{u}}$ of $\bar{u}$ and a positive constant ϵ such that for $0 < \delta_i \leq \epsilon$,

$$\Phi_{\delta_i}(u, \bar{t}) = 0$$

has a unique solution in $w_{\bar{u}}, i = 1, 2, \cdots$.

Proof. For given $(\bar{u}, \bar{t}) \in H^{-1}(0)$, Theorem 2.5 leads to that there are an open neighborhood $w_{\bar{u}}$ of $\bar{u}$ and a positive constant ϵ such that for $u \in w_{\bar{u}}$ and $0 < \delta_i \leq \epsilon$,

$$\text{sgn} \det(\Phi_{\delta_i})_u'(u, \bar{t}) = \text{sgn} \det H_u'(\bar{u}, \bar{t}).$$

We can choose $w_{\bar{u}}$ such that $\bar{u}$ is the unique solution of $H(u, \bar{t}) = 0$ in $\overline{w}_{\bar{u}}$. Then

$$\deg(H(\cdot, \bar{t}), \overline{w}_{\bar{u}}, 0) = \text{sgn} \det H_u'(\bar{u}, \bar{t}).$$

Since Φ_{δ_i} tends to H as i approaches ∞, for δ_i small enough,

$$\deg(H(\cdot, \bar{t}), \overline{w}_{\bar{u}}, 0) = \deg(\Phi_{\delta_i}(\cdot, \bar{t}), \overline{w}_{\bar{u}}, 0).$$

Moreover, since 0 is a regular value of Φ_{δ_i},

$$\deg(\Phi_{\delta_i}(\cdot, \bar{t}), \overline{w}_{\bar{u}}, 0) = \sum \text{sgn} \det(\Phi_{\delta_i})_u'(u, \bar{t}),$$

where the sum $\sum$ runs through $\{u \in \overline{w}_{\bar{u}} : \Phi_{\delta_i}(u, \bar{t}) = 0\}$.

In summary,

$$\begin{aligned}
\text{sgn} \det H_u'(\bar{u}, \bar{t}) &= \deg(H(\cdot, \bar{t}), \overline{w}_{\bar{u}}, 0) \\
&= \deg(\Phi_{\delta_i}(\cdot, \bar{t}), \overline{w}_{\bar{u}}, 0) \\
&= \sum \text{sgn} \det(\Phi_{\delta_i})_u'(u, \bar{t}) \\
&= \sum \text{sgn} \det H_u'(\bar{u}, \bar{t}).
\end{aligned}$$

The last sum only contains one term, that is, there is a unique $u \in w_{\bar{u}}$ such that

$$\Phi_{\delta_i}(u, \bar{t}) = 0. \P$$

Similar to the regularity of a sequence of simplicial triangulations, we can introduce the regularity of a simplicial triangulation of an unbounded underlying convex set. This regularity is, of course, based on certain uniformity of the diameters of simplices of the triangulation.

Next theorem provides a standard method to find the starting simplex for simplicial homotopy algorithms.

Theorem 2.8 *For a given positive number δ, let $\mathbf{T}^\delta$ be a triangulation of $\mathbf{R}^m \times [0,1]$ such that the restriction of $\mathbf{T}^\delta$ on $\mathbf{R}^m \times \{0\}$ is regular. If $(\bar{u},0) \in H_0^{-1}(0)$ is a barycenter of some m-dimensional simplex $\sigma_{\bar{u}}^m$, $w_{\bar{u}}$ is a neighborhood of $\bar{u}$ with $w_{\bar{u}} \times \{0\} \cap H_0^{-1}(0) = \{(\bar{u},0)\}$, and $H_u'(\bar{u},0)$ is nonsingular, then there is a positive number ϵ such that for $0 < \delta \leq \epsilon$, $\sigma_{\bar{u}}^m$ is the unique simplex in $w_{\bar{u}}$ containing a zero of $\Phi_\delta(\cdot,0)$.*

Proof. Let $(\bar{u},0) \in H_0^{-1}(0)$ be the barycenter of the simplex $\sigma_{\bar{u}}^m$. Since $\mathbf{T}^\delta$ is regular, there is a positive number α independent of δ such that

$$\alpha\delta \leq \frac{V\sigma_{\bar{u}}^m}{\delta^{m-1}} \leq \frac{h\sigma_{\bar{u}}^m}{m},$$

where $V\sigma_{\bar{u}}^m$ and $h\sigma_{\bar{u}}^m$ denote respectively the volume and the smallest height of the simplex $\sigma_{\bar{u}}^m$. Thus,

$$\{u : |u - \bar{u}| < \alpha\delta\} \subset \text{Int}\sigma_{\bar{u}}^m.$$

Since $H_u'(\bar{u},0)$ is nonsingular, there is a $\beta > 0$ such that

$$|H_u'(\bar{u},0)u| \geq \beta$$

for all $|u| = 1$. Let $\tilde{w} \subset w_{\bar{u}}$ be an open neighborhood of $(\bar{u},0)$. For any $u, \tilde{u} \in \tilde{w}$ with $u \neq \tilde{u}$,

$$\frac{|R(|u - \tilde{u}|)|}{|u - \tilde{u}|} < \frac{\alpha\beta}{\alpha + 1},$$

where R is the remainder of Taylor formula, that is,

$$H(u,0) = H(\tilde{u},0) + H_u'(\tilde{u},0)(u - \tilde{u}) + R(|u - \tilde{u}|).$$

Since $\bar{u}$ is the unique solution of $H(u,0) = 0$ in $w_{\bar{u}}$. There is a positive number ϵ such that for $\delta \in (0,\epsilon]$, any simplex in $w_{\bar{u}}$ containing a solution of $\Phi_\delta(u,0) = 0$ must lie in $\tilde{w}$. Now, we need only to prove that any simplex $\sigma^m = \{u^j\}_0^m \neq \sigma_{\bar{u}}^m$ doesn't contain any zeros of $\Phi_\delta(\cdot,0)$.

For every $\sum_{j=0}^m \lambda_j u^j \in \sigma^m$ with $\sum_{j=0}^m \lambda_j = 1$ and $\lambda_j \geq 0$,

$$\left| H\left(\sum_{j=0}^m \lambda_j u^j, 0\right) \right| = \left| H\left(\sum_{j=0}^m \lambda_j u^j, 0\right) - H(\bar{u},0) \right|$$

$$= \left| H_u'(\bar{u},0)\left(\sum_{j=0}^m \lambda_j u^j - \bar{u}\right) + R\left(\left|\sum_{j=0}^m \lambda_j u^j - \bar{u}\right|\right) \right|$$

$$\geq \left| H_u'(\bar{u},0)\left(\sum_{j=0}^m \lambda_j u^j - \bar{u}\right) \right| - \left| R\left(\left|\sum_{j=0}^m \lambda_j u^j - \bar{u}\right|\right) \right|$$

$$> \beta \left| \sum_{j=0}^{m} \lambda_j u^j - \overline{u} \right| - \frac{\alpha\beta}{\alpha+1} \left| \sum_{j=0}^{m} \lambda_j u^j - \overline{u} \right|$$

$$\geq \frac{\beta}{\alpha+1} \left| \sum_{j=0}^{m} \lambda_j u^j - \overline{u} \right|$$

$$\geq \frac{\alpha\beta\delta}{\alpha+1}.$$

Similarly,

$$\left| \sum_{j=0}^{m} \lambda_j H(u^j, 0) - H\left(\sum_{j=0}^{m} \lambda_j u^j, 0 \right) \right|$$

$$= \left| \sum_{j=0}^{m} \lambda_j \left(H(u^j, 0) - H\left(\sum_{j=0}^{m} \lambda_j u^j, 0 \right) \right) \right|$$

$$= \left| \sum_{j=0}^{m} \lambda_j \left[H'_u\left(\sum_{j=0}^{m} \lambda_j u^j, 0 \right) \left(u^j - \sum_{j=0}^{m} \lambda_j u^j \right) \right. \right.$$

$$\left. \left. + R\left(\left| u^j - \sum_{j=0}^{m} \lambda_j u^j \right| \right) \right] \right|$$

$$= \left| \sum_{j=0}^{m} \lambda_j R\left(\left| u^j - \sum_{j=0}^{m} \lambda_j u^j \right| \right) \right|$$

$$\leq \sum_{j=0}^{m} \lambda_j \left| R\left(\left| u^j - \sum_{j=0}^{m} \lambda_j u^j \right| \right) \right|$$

$$\leq \sum_{j=0}^{m} \lambda_j \frac{\alpha\beta}{\alpha+1} \left| u^j - \sum_{j=0}^{m} \lambda_j u^j \right|$$

$$\leq \sum_{j=0}^{m} \lambda_j \frac{\alpha\beta}{\alpha+1} \delta$$

$$= \frac{\alpha\beta\delta}{\alpha+1}.$$

Thus,

$$\left| \Phi_\delta\left(\sum_{j=0}^{m} \lambda_j u^j, 0 \right) \right| = \left| \sum_{j=0}^{m} \lambda_j H(u^j, 0) \right|$$

$$\geq \left| H\left(\sum_{j=0}^{m} \lambda_j u^j, 0 \right) \right|$$

$$-\left|\sum_{j=0}^{m}\lambda_j H(u^j,0) - H\left(\sum_{j=0}^{m}\lambda_j u^j,0\right)\right|$$

$$> \frac{\alpha\beta\delta}{\alpha+1} - \frac{\alpha\beta\delta}{\alpha+1} = 0.$$

By Theorem 6.2.12, for ϵ small enough,

$$\deg(\Phi_\delta(\cdot,0),\tilde{w},0) = \deg(H(\cdot,0),\tilde{w},0) \neq 0.$$

Then Theorem 6.2.14 guarantees that $\Phi_\delta(u,0) = 0$ has at least one solution in $\tilde{w}$. The result thus follows. ¶

Remark 2.9 With the argument similar to the proof of Theorem 2.8, we can prove $|\Phi_\delta|_{\partial\sigma_{\overline{u}}^m}| > 0$, that is, every zero of Φ_δ in $w_{\overline{u}}$ must lie in $\mathrm{Int}\sigma_{\overline{u}}^m$.

Suppose that $H_0^{-1}(0) = \{(\overline{u}^1,0),\cdots,(\overline{u}^l,0)\}$ and every $(\overline{u}^i,0)$ is the barycenter of some m-dimensional simplex $\sigma_{\overline{u}^i}^m$. If $w_{\overline{u}^i}$ is a neighborhood of $(\overline{u}^i,0)$ with $w_{\overline{u}^i} \cap H_0^{-1}(0) = \{(\overline{u}^i,0)\}$, and $H'_u(\overline{u}^i,0)$ is nonsingular then there is a positive number ϵ such that for $0 < \delta \leq \epsilon$, $\sigma_{\overline{u}^i}^m$ is the unique simplex in $w_{\overline{u}^i}$ containing the zero of $\Phi_\delta(\cdot,0)$, $i = 1,\cdots,l$.

Remark 2.10 With Theorems 2.7, 2.1 and 2.5, one can prove Index Theorem 1.8.

Now, we introduce the concept of refining triangulations. For our purpose of focusing only on the basic idea of simplicial homotopy algorithms, the following is in fact a quite special kind of refining simplicial triangulations.

Definition 2.11 A triangulation $\mathbf{T}^\delta$ of $\mathbf{R}^m \times [0,1)$ is called a refining simplicial triangulation if it satisfies

(1) All vertices of the triangulation lie on the levels $\mathbf{R}_k^m = \mathbf{R}^m \times \{1 - 2^{-k}\}$, $k = 0,1,2,\cdots$;

(2) Every simplex in the triangulation lies between the levels $\mathbf{R}_k^m$ and $\mathbf{R}_{k+1}^m$ for some $k \in \{0,1,2,\cdots,\}$;

(3) The projective diameter on $\mathbf{R}_k^m$ of every simplex lying between the levels $\mathbf{R}_k^m$ and $\mathbf{R}_{k+1}^m$ is at most $\sqrt{m} \cdot 2^{-k}\delta$, where $\delta > 0$ is the mesh parameter of $\mathbf{T}^\delta$ (the maximum diameter of simplices in the triangulation).

For example, the J_3 triangulation of $\mathbf{R}^m \times [0,1)$ is a refining simplicial triangulation. Fig. 8.13 illustrates the J_3 triangulation when $m = 1$. (cf. [Todd, 1976])

Definition 2.12 Let $H : \mathbf{R}^m \times [0,1] \to \mathbf{R}^m$ be differentiable and $\mathbf{T}^\delta$ be a refining simplicial triangulation of $\mathbf{R}^m \times [0,1)$ with δ as its mesh parameter. Φ_δ is called a piecewise linear approximation of H with respect to $\mathbf{T}^\delta$ if

$$\Phi_\delta(u,t) = \begin{cases} H(u,t), & \text{if } (u,t) \in \mathbf{R}^m \times \{1\} \text{ or} \\ & (u,t) \text{ is a vertex of } \mathbf{T}^\delta, \\ \sum_{j=0}^{m+1} \lambda_j \Phi_\delta(u^j, t_j) = \sum_{j=0}^{m+1} \lambda_j H(u^j, t_j), \\ \quad \text{if } (u,t) = \sum_{j=0}^{m+1} \lambda_j(u^j, t_j) \in \sigma^{m+1} \\ \quad = \{(u^j, t_j)\}_0^{m+1} \in \mathbf{T}^\delta \\ \quad \text{with } \lambda_j \geq 0 \text{ and } \sum_{j=0}^{m+1} \lambda_j = 1. \end{cases}$$

It is clear that Φ_δ is continuous.

Definition 2.13 Let $\{\mathbf{T}^{\delta_i}\}$ be a sequence of refining simplicial triangulations of $\mathbf{R}^m \times [0,1)$ with their mesh parameters $\delta_i \to 0$ as $i \to \infty$. $\{\mathbf{T}^{\delta_i}\}$ is called a sequence of regular refining simplicial triangulations if for some given positive number α and all $(m+1)$-dimensional simplices $\sigma = \{x^i\}_0^{m+1}$ in $\mathbf{T}^{\delta_i}$, $i = 1, 2, \cdots$, the volume of

$$\left\{ x \in \mathbf{R}^{m+1} : x = x^0 + \sum_{i=0}^{m+1} \lambda_i(x^i - x^0), 0 \leq \lambda_i \leq 1 \right\}$$

satisfies

$$\det \begin{pmatrix} x^0 & \cdots & x^{m+1} \\ 1 & \cdots & 1 \end{pmatrix} = \det(x^1 - x^0, \cdots, x^{m+1} - x^0) \geq \alpha \delta_\sigma^{m+1},$$

where δ_σ denotes the diameter of σ.

It is easy to see that the sequence of the triangulations $\{\delta_i J_3\}$ of $\mathbf{R}^m \times [0,1)$ is regular.

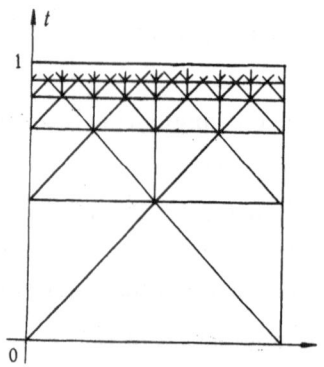

Figure 8.13

Remark 2.14   The results similar to Theorems 2.5 and 2.7 are also true for a sequence of regular refining simplicial triangulations. In this case, for every differentiable bounded path L in $H^{-1}(0)$, there are a positive number ϵ and a neighborhood w of the path such that for $0 < \delta_i \le \epsilon$, there is a unique piecewise linear path $L_{\Phi_{\delta_i}}$ of $\Phi_{\delta_i}^{-1}(0)$ in w, and the results of Theorems 2.5 and 2.7 are still true for Φ_{δ_i}. Let $L_{\Phi_{\delta_i}} = (u(\theta), t(\theta))$ for $0 \le \theta \le \theta_1$ and $\lim_{\theta \to \theta_1} t(\theta) = 1$, $\lim_{\theta \to \theta_1} u(\theta) = u^*$. Then

$$0 = \lim_{\theta \to \theta_1} \Phi_{\delta_i}(u(\theta), t(\theta)) = \Phi_{\delta_i}(u^*, 1),$$

that is, u^* is a zero of the mapping $H(\cdot, 1)$.

§3. PL Homotopy Algorithms Work with Probability one

This section is devoted to discussion of the problems of finding zeros of polynomial mapping $P : \mathbf{C}^n \to \mathbf{C}^n$. Let $P(z) = (P_1(z), \cdots, P_n(z))$ be a polynomial mapping and $q_i \ge 1$ be the degree of P_j, $j = 1, \cdots, n$. We choose the polynomial mapping $Q : \mathbf{C}^n \to \mathbf{C}^n$ with its j-th component $Q_j(z) = z_j^{q_j} - b_j^{q_j}$, $b_j \ne 0$. It is clear that Q has the following $q = \prod_{j=1}^n q_j$ zeros:

$$(b_1 e^{i \frac{2k_1 \pi}{q_1}}, \cdots, b_n e^{i \frac{2k_n \pi}{q_n}}), \quad k_j = 1, \cdots, q_j, j = 1, \cdots, n.$$

Let $H : \mathbf{C}^n \times [0, 1] \to \mathbf{C}^n$ be the linear homotopy between Q and P with $H(z, 0) = Q(z)$ and $H(z, 1) = P(z)$, that is,

$$H(z, t) = tP(z) + (1 - t)Q(z).$$

Consider the zero set

$$H^{-1}(0) = \{(z, t) \in \mathbf{C}^n \times [0, 1] : H(z, t) = 0\}.$$

Theorem 7.1.13 says that for almost every polynomial mapping $P : \mathbf{C}^n \to \mathbf{C}^n$, $H^{-1}(0)$ consists of $q = \prod_{j=1}^n q_j$ 1-dimensional differentiable connected paths, and every path links a zero of Q and a zero of P and the parameter t is always strictly increasing as following each of the path from the zeros of Q to the zeros of P.

Now, the zeros of Q are known. Following the paths of $H^{-1}(0)$ from 0 to 1, one can arrive at all zeros of P. But for every fixed $t \in [0, 1]$, $H_t(z) = tP(z) + (1 - t)Q(z)$ is a polynomial mapping. Generally speaking, to locate directly the all zeros of that mapping is hardly possible as long as $t \neq 0$.

Let $\mathbf{T}$ be a triangulation of $\mathbf{C}^n \times [0, 1] = \mathbf{R}^{2n} \times [0, 1]$ with mesh size δ. Let further Φ_δ be the piecewise linear approximation of H with respect to $\mathbf{T}$. Since Φ_δ is affine on every simplex of the triangulation, it is easy, at least in theory, to locate the zeros of Φ_δ on a simplex. In fact, it need only to solve a linear system of equations. Furthermore, if 0 is a regular value of Φ_δ and the zero set of Φ_δ intersects with some $(2n + 1)$-dimensional simplex, then the intersection is a line segment with two ends lying in the interiors of two different $2n$-dimensional faces of the simplex. Thus,

$$\Phi_\delta^{-1}(0) = \{(z, t) \in \mathbf{C}^n \times [0, 1] : \Phi_\delta(z, t) = 0\},$$

the zero set of Φ_δ, consists of some piecewise linear paths and loops not intersecting each other.

In Section 2, we have shown that the feasibility of simplicial homotopy algorithms is dependent on the fact if the piecewise linear paths intersect with the v-dimensional skeleton $\mathbf{T}^v$ of the triangulation for at least one of $v = 0, 1, \cdots, 2n - 1$. In fact, if some path intersects with $\mathbf{T}^v$ for some $v \in \{0, 1, \cdots, 2n - 1\}$, then after some pivots, the computation will arrive at a v-dimensional face from some $(2n + 1)$-dimensional simplex. But there are more than one $(2n + 1)$-dimensional simplices sharing the v-dimensional face with the current $(2n+1)$-dimensional simplex. Then the complementary pivoting algorithm fails. But suppose that the paths don't intersect with $\mathbf{T}^v$ for all $v = 0, 1, \cdots, 2n - 1$, when the computation passes a $(2n + 1)$-dimensional simplex, the complementary pivoting procedure determines uniquely a new $2n$-dimensional face (an exit) from the previous $2n$-dimensional face (an entrance) of the simplex. In this case, there is a unique $(2n + 1)$-dimensional simplex sharing the new $2n$-face with the simplex. Thus, the algorithm can continue in the new $(2n+1)$-dimensional simplex.

In this section, we will prove that for almost every polynomial mapping, the zero set of Φ_δ doesn't intersect with $\mathbf{T}^v$ for all $v = 0, 1, 2, \cdots, 2n - 1$. In this sense, we say that the simplicial homotopy complementary pivoting algorithm for polynomial mappings is feasible with probability one.

Let

$$\mathbf{T}_0^v = \{\sigma \in \mathbf{T}^v : \sigma \subset \mathbf{C}^n \times \{0\}\}$$

and

$$\mathbf{T}_+^v = \{\sigma \in \mathbf{T}^v : \sigma \subset \mathbf{C}^n \times (0, 1]\}.$$

It is clear that $\mathbf{T}^v = \mathbf{T}_0^v \bigcup \mathbf{T}_+^v$ for $v = 0, 1, 2, \cdots, 2n+1$ and $\mathbf{T}_0^{2n+1} = \emptyset$.

First, we discuss the properties of Φ_δ at the starting level $\mathbf{C}^n \times \{0\}$.

Theorem 3.1   *Let* $\mathbf{T}$ *be a triangulation of* $\mathbf{C}^n \times [0, 1]$ *with mesh size* δ*. Then for almost every* $b \in \mathbf{C}^n$,

$$H^{-1}(0) \bigcap \mathbf{T}_0^v = \emptyset$$

and for all $v = 0, 1, \cdots, 2n - 1$

$$\Phi_\delta^{-1}(0) \bigcap \mathbf{T}_0^v = \emptyset.$$

Proof.   Define $\alpha : \mathbf{C}^n \times \{0\} \to \mathbf{C}^n$ by

$$\alpha(z, 0) = (z_1^{q_1}, \cdots, z_n^{q_n}) \text{ for } (z, 0) \in \mathbf{C}^n \times \{0\}.$$

Let β be the piecewise linear approximation of α with respect to the restriction of the triangulation $\mathbf{T}$ on $\mathbf{C}^n \times \{0\}$.

Let $\mathbf{T}_0^v = \bigcup_{k=1}^\infty \sigma_k^v, v = 0, 1, 2, \cdots, 2n - 1$, where σ_k^v is an open v-dimensional simplex of T on $\mathbf{C}^n \times \{0\}$. Consider the restrictions of the mappings α and β on $\mathbf{T}_0^v$, $v = 0, 1, \cdots, 2n - 1$. When $v < 2n$, by Lemma 7.1.2, the measure of the image set $\text{Image}(\alpha|_{\sigma_k^v})$ is zero in $\mathbf{C}^n = \mathbf{R}^{2n}$. Thus, the measure of $\text{Image}(\alpha|_{\mathbf{T}_0^v})$ is also zero in $\mathbf{C}^n$. Similarly, the measure of $\text{Image}(\beta|_{\mathbf{T}_0^v})$ is zero in $\mathbf{C}^n$. Then we obtain that both the measures of

$$\text{Image}(\alpha|_{\mathbf{T}_0^0 \bigcup \mathbf{T}_0^1 \bigcup \mathbf{T}_0^2 \bigcup \cdots \bigcup \mathbf{T}_0^{2n-1}})$$

and of

$$\text{Image}(\beta|_{\mathbf{T}_0^0 \bigcup \mathbf{T}_0^1 \bigcup \mathbf{T}_0^2 \bigcup \cdots \bigcup \mathbf{T}_0^{2n-1}})$$

are zero in $\mathbf{C}^n$, that is, both the measures of

$$\text{Image}(\alpha|_{\overline{\mathbf{T}_0^{2n-1}}})$$

and of

$$\text{Image}(\beta|_{\overline{\mathbf{T}_0^{2n-1}}})$$

are zero in $\mathbf{C}^n$. Thus, we can choose a $c \in \mathbf{C}^n$ such that $c_j \neq 0$ for $j = 1, \cdots, n$, and

$$c \notin \text{Image}(\alpha|_{\overline{\mathbf{T}_0^{2n-1}}})$$

and

$$c \notin \text{Image}(\beta|_{\overline{\mathbf{T}_0^{2n-1}}}).$$

For $j = 1, \cdots, n$, choose b_j such that $b_j^{q_j} = c_j$. Define $Q : \mathbf{C}^n \to \mathbf{C}^n$ by

$$Q(z) = (z_1^{q_1} - b_1^{q_1}, \cdots, z_n^{q_n} - b_n^{q_n}).$$

Let the homotopy $H(z,t) = tP(z) + (1-t)Q(z)$ and Φ_δ be the piecewise linear approximation of H with respect to T.

By the choice of the c, we know that for almost every $b \in \mathbf{C}^n$,

$$H^{-1}(0) \bigcap \mathbf{T}_0^v = \emptyset$$

and for all $v = 0, 1, \cdots, 2n - 1$

$$\Phi_\delta^{-1}(0) \bigcap \mathbf{T}_0^v = \emptyset. \P$$

Now we prove the two main results of this section.

Theorem 3.2   *Let* $\mathbf{T}$ *be a triangulation of* $\mathbf{C}^n \times [0,1]$ *and* $Q(z) = (Q_1(z), \cdots, Q_n(z))$ *with* $Q_j(z) = z_j^{q_j} - b_j^{q_j}$. *If for some* $b \in \mathbf{C}^n$,

$$H^{-1}(0) \bigcap \mathbf{T}_0^v = \emptyset \ for \ v = 0, 1, 2, \cdots, 2n - 1,$$

then for almost every polynomial mapping $P : \mathbf{C}^n \to \mathbf{C}^n$,

$$H^{-1}(0) \bigcap \mathbf{T}^v = \emptyset \ for \ all \ v = 0, 1, \cdots, 2n - 1.$$

Proof.   Let $P : \mathbf{C}^n \to \mathbf{C}^n$ be a polynomial mapping and $q_j > 0$ be the degree of its j-th component P_j for $j = 1, \cdots, n$. Rewrite $P(\cdot)$ as $P(\cdot; \omega)$ in the way as in the previous chapter, where $\omega \in \mathbf{C}^N$ is the coefficients of the polynomial mapping. Let $\omega = (a, c)$, where c is the constant term of the polynomial mapping.

Define $\mu : \mathbf{C}^n \times (0,1] \times \mathbf{C}^{N-n} \to \mathbf{C}^{N-n} \times \mathbf{C}^n$ by

$$\mu(z, t; a) = \left(a, \frac{(t-1)Q(z) - tP(z; a, 0)}{t} \right).$$

It is clear that μ is analytic in z and a. It is also easy to see that for $t > 0$, (z, t) is a zero of the linear homotopy $H(\cdot, \cdot; a, c)$ between $P(\cdot; a, c)$ and $Q(\cdot)$ if and only if

$$\mu(z, t; a) = (a, c).$$

Let $\mathbf{T}_+^v = \sum_{k=1}^\infty \sigma_k^v$, where σ_k^v denotes an open v-dimensional simplex of $\mathbf{T}$ with $t > 0$, $v = 0, 1, \cdots, 2n - 1$; $k = 1, 2, \cdots$.

Consider the restriction of μ on $\mathbf{C}^{N-n} \times \sigma_k^v$. Since $(z, t; a) \in \sigma_k^v \times \mathbf{C}^{N-n}$ if and only if there are unique $\lambda_0, \lambda_1, \cdots, \lambda_v > 0$ with $\sum_{j=0}^{v} \lambda_j = 1$ such that

$$(z, t; a) = \left(\sum_{j=0}^{v} \lambda_j e^j, a \right),$$

where $e^0, e^1, \cdots, e^v$ are the all vertices of the simplex σ_k^v. Then

$$\mu|_{\sigma_k^v \times \mathbf{C}^{N-n}} : \sigma_k^v \times \mathbf{C}^{N-n} \to \mathbf{C}^{N-n} \times \mathbf{C}^n$$

can be rewritten as

$$\mu(\lambda_1, \cdots, \lambda_v; a) = \left(a, \frac{(t-1)Q(z) - tP(z; a, 0)}{t} \right).$$

By Lemma 7.1.2, the measure of the image $\mathrm{Image}(\mu|_{\sigma_k^v \times \mathbf{C}^{N-n}})$ is zero in $\mathbf{C}^{N-n} \times \mathbf{C}^n$, so is the measure of $\mathrm{Image}(\mu|_{\mathbf{T}_+^v \times \mathbf{C}^{N-n}})$.

We have shown that for $t > 0$, (z, t) is a zero of $H(\cdot, \cdot; a, c)$ if and only if $\mu(z, t; a) = (a, c)$. Thus, $H(\cdot, \cdot; a, c)$ has a zero in $\mathbf{T}_+^v$ if and only if $(a, c) \in \mathrm{Image}(\mu|_{\mathbf{T}_+^v \times \mathbf{C}^{N-n}})$. But the measure of $\mathrm{Image}(\mu|_{\mathbf{T}_+^v \times \mathbf{C}^{N-n}})$ is zero in $\mathbf{C}^{N-n} \times \mathbf{C}^n$. This means that for almost every $(a, c) \in \mathbf{C}^{N-n} \times \mathbf{C}^n$ (i.e., for almost every polynomial mapping $P(z; a, c)$),

$$H^{-1}(0) \bigcap \mathbf{T}_+^v = \emptyset.$$

Then the hypothesis $H^{-1}(0) \bigcap \mathbf{T}_0^v = \emptyset$ leads to

$$H^{-1}(0) \bigcap \mathbf{T}^v = \emptyset \text{ for all } v = 0, 1, \cdots, 2n - 1. ¶$$

Theorem 3.3 *Let $\mathbf{T}$ be a triangulation of $\mathbf{C}^n \times [0, 1]$ and $Q(z) = (Q_1(z), \cdots, Q_n(z))$ with $Q_j(z) = z_j^{q_j} - b_j^{q_j}$, $b_j \neq 0$. If for some $b \in \mathbf{C}^n$,*

$$\Phi_\delta^{-1}(0) \bigcap \mathbf{T}_0^v = \emptyset \text{ for all } v = 0, 1, 2, \cdots, 2n - 1,$$

then for almost every polynomial mapping $P : \mathbf{C}^n \to \mathbf{C}^n$,

$$\Phi_\delta^{-1}(0) \bigcap \mathbf{T}^v = \emptyset \text{ for all } v = 0, 1, \cdots, 2n - 1.$$

Proof. Define $\mu : \mathbf{C}^n \times (0, 1) \times \mathbf{C}^{N-n} \to \mathbf{C}^{N-n} \times \mathbf{C}^n$ as follows: For every $(z, t; a) \in \mathbf{C}^n \times (0, 1] \times \mathbf{C}^{N-n}$, there is a unique open simplex $\sigma^v = \{(z^j, t_j)\}_0^v$ of $\mathbf{T}$ such that (z, t) is an interior point of the simplex, where $(z^0, t_0), \cdots, (z^v, t_v)$ are the all vertices of σ^v. Then (z, t) can be uniquely represented as

$$(z, t) = \sum_{j=0}^{v} \lambda_j(z^j, t_j), \quad \lambda_0 > 0, \cdots, \lambda_v > 0, \quad \sum_{j=0}^{v} \lambda_j = 1.$$

Let

$$\mu(z, t; a) = \left(a, \frac{1}{\sum_{j=0}^{v} \lambda_j t_j} \sum_{j=0}^{v} \lambda_j[(t_j - 1)Q(z^j) - t_j P(z^j; a, 0)] \right).$$

It is clear that μ is well defined and is a piecewise linear mapping.

Now consider the zero set of Φ_δ. We have shown that there are zeros of Φ_δ in σ^v is equivalent to that there are $\lambda_0, \cdots, \lambda_v > 0$ with $\sum_{j=0}^{v} \lambda_j = 1$ such that

$$\Phi_\delta \left(\sum_{j=0}^{v} \lambda_j(z^j, t_j) \right)$$

$$= \sum_{j=0}^{v} \lambda_j \Phi_\delta(z^j, t_j)$$

$$= \sum_{j=0}^{v} \lambda_j H(z^j, t_j; a, c)$$

$$= \sum_{j=0}^{v} \lambda_j[t_j P(z^j; a, c) + (1 - t_j)Q(z^j)]$$

$$= \sum_{j=0}^{v} \lambda_j[t_j(P(z^j; a, 0) + c) + (1 - t_j)Q(z^j)]$$

$$= c \sum_{j=0}^{v} \lambda_j t_j - \sum_{j=0}^{v} \lambda_j[(t_j - 1)Q(z^j) - t_j(P(z^j; a, 0)]$$

$$= 0.$$

It is also equivalent to that there are $\lambda_0, \cdots, \lambda_v > 0$ with $\sum_{j=0}^{v} \lambda_j = 1$ such that

$$(a, c) = \mu(z, t; a).$$

Therefore, $\Phi_\delta^{-1}(0) \bigcap \sigma^v \neq \emptyset$ if and only if $(a, c) \in \text{Image}(\mu|_{\mathbf{C}^{N-n} \times \sigma^v})$.

Let $\mathbf{T}_+^v = \bigcup_{k=1}^{\infty} \sigma_k^v$. For a σ_k^v, let $\sigma_k^v = \{(z^j, t_j)\}_0^v$. Then

$$\mu|_{\sigma_k^v \times \mathbf{C}^{N-n}} : \sigma_k^v \times \mathbf{C}^{N-n} \to \mathbf{C}^{N-n} \times \mathbf{C}^n$$

can be rewritten as

$$\mu(\lambda_1, \cdots, \lambda_v; a) = \left(a, \frac{1}{\sum_{j=0}^{v} \lambda_j t_j} \sum_{j=0}^{v} \lambda_j [(t_j - 1)Q(z^j) - t_j P(z^j; a, 0)]\right),$$

where $\lambda_0 = 1 - \sum_{j=1}^{v} \lambda_j$. Lemma 7.1.2 leads to that the measure of the set Image$(\mu|_{\sigma_k^v \times \mathbf{C}^{N-n}})$ is zero in $\mathbf{C}^{N-n} \times \mathbf{C}^n$, and thus the measure of the set Image$(\mu|_{\mathbf{T}_+^v \times \mathbf{C}^{N-n}})$ is also zero in $\mathbf{C}^{N-n} \times \mathbf{C}^n$. It leads to that for almost every $(a, c) \in \mathbf{C}^{N-n} \times \mathbf{C}^n$ (i.e., for almost every polynomial mapping), $\Phi_\delta^{-1}(0) \cap \mathbf{T}_+^v = \emptyset$. Then the hypothesis $\Phi_\delta^{-1}(0) \cap \mathbf{T}_0^v = \emptyset$ gives that for all $v = 0, 1, \cdots, 2n - 1$

$$\Phi_\delta^{-1}(0) \cap \mathbf{T}^v = \emptyset. \P$$

Notice that the measure of the union of a countable sequence of sets with zero measure is also zero. We can extend Theorems 3.1, 3.2 and 3.3 to the case of a sequence of triangulations

$$\{\mathbf{T}^{\delta_i} : i = 1, 2, \cdots\}$$

of $\mathbf{C}^n \times [0, 1]$ and obtain the following three theorems.

Theorem 3.4 Let $\{\mathbf{T}^{\delta_i} : i = 1, 2, \cdots\}$ be a sequence of simplicial triangulations of $\mathbf{R}^n \times [0, 1]$ and δ_i be the mesh size of $\mathbf{T}^{\delta_i}$. Then for almost every $b \in \mathbf{C}^n$,

$$H^{-1}(0) \cap (\mathbf{T}^{\delta_i})_0^v = \emptyset$$

and for all $i = 1, 2, \cdots$ and $v = 0, 1, \cdots, 2n - 1$

$$\Phi_{\delta_i}^{-1}(0) \cap (\mathbf{T}^{\delta_i})_0^v = \emptyset. \P$$

Theorem 3.5 Let $\{\mathbf{T}^{\delta_i} : i = 1, 2, \cdots\}$ be a sequence of triangulations of $\mathbf{C}^n \times [0, 1]$ and δ_i be the mesh size of $\mathbf{T}^{\delta_i}$. Let further $Q(z) = (Q_1(z), \cdots, Q_n(z))$ with $Q_j(z) = z_j^{q_j} - b_j^{q_j}$. If for some $b \in \mathbf{C}^n$,

$$H^{-1}(0) \cap (\mathbf{T}^{\delta_i})_0^v = \emptyset \text{ for all } i = 1, 2, \cdots, \quad v = 0, 1, 2, \cdots, 2n - 1,$$

then for almost every polynomial mapping $P : \mathbf{C}^n \to \mathbf{C}^n$,

$$H^{-1}(0) \cap (\mathbf{T}^{\delta_i})^v = \emptyset \text{ for all } i = 1, 2, \cdots, \quad v = 0, 1, \cdots, 2n - 1. \P$$

Theorem 3.6 Let $\{\mathbf{T}^{\delta_i} : i = 1, 2, \cdots\}$ be a sequence of triangulations of $\mathbf{C}^n \times [0, 1]$ and δ_i be the mesh size of $\mathbf{T}^{\delta_i}$. Let further $Q(z) = (Q_1(z), \cdots, Q_n(z))$ with $Q_j(z) = z_j^{q_j} - b_j^{q_j}$. If for some $b \in \mathbf{C}^n$,

$$\Phi_{\delta_i}^{-1}(0) \cap (\mathbf{T}^{\delta_i})_0^v = \emptyset \text{ for all } i = 1, 2, \cdots, \quad v = 0, 1, 2, \cdots, 2n - 1,$$

then for almost every polynomial mapping $P : \mathbf{C}^n \to \mathbf{C}^n$,

$$\Phi_{\delta_i}^{-1}(0) \bigcap (\mathbf{T}^{\delta_i})^v = \emptyset \text{ for all } i = 1, 2, \cdots, \quad v = 0, 1, \cdots, 2n - 1.\P$$

Remark 3.7   In Theorems 3.1, 3.2 and 3.3, if we substitute a refining simplicial triangulation for the simplicial triangulation, the results are also true. In this case, let Φ_δ be the piecewise linear approximation of H with respect to the refining simplicial triangulation $\mathbf{T}^\delta$ with mesh parameter δ. Then every connected component of the zero set $\Phi_\delta^{-1}(0)$ with only one end in $\mathbf{C}^n \times \{0\}$ is an infinite piecewise linear path.

Remark 3.8   As for the implementation of the algorithms discussed in this chapter, we usually employ the complementary pivoting procedure to follow the connected component of the zero set $\Phi_\delta^{-1}(0)$. In this case, the labelling rules of the vertices of the triangulations are important. There are two kinds of labellings, one is the integer labelling and the other is the vector labelling. When the pivoting computation follows the zero set $\Phi_\delta^{-1}(0)$ with the integer labelling, the computational sequences of simplices are near some above-mentioned finite or infinite piecewise linear paths. When the vector-labelling is employed, the elements of the methods is the so-called complete simplices, and the pivoting computation is implemented by using lexicographic systems. See [Todd, 1976] or [Wang, 1986] for a detailed discussion. In this case, the above-mentioned finite or infinite paths are contained inside the computational sequences of simplices.

Remark 3.9   Based on [Wang, 1991], [Wang, 1990] explores the two-dimensional structure of the paths of simplicial or PL (piecewise linear) homotopy methods when vector labelling is employed. The paths are sequences of wing-shaped pieces of ruled surfaces. It leads geometrically to an improvement of the feasibility of the vector-labelling PL homotopy methods in the following sense: Once a so-called complete simplex is found somehow , the complementary pivoting computation will continue along a sequence of successively adjacent complete simplices forever with no possibility of branching until an expected simplex appears. Thus in the terminology of Theorem 3.3, $\Phi_\delta^{-1}(0) \bigcap \mathbf{T}^v = \emptyset$ for all $v = 0, 1, \cdots, 2n - 1$, is now not a prerequisite condition for the feasibility of PL homotopy methods any more.

References

[1] Abraham, R. & Robbins, J., *Transversal mappings and flows*, Benjamin, New York, 1967.

[2] Allgower, E. L. & Georg, K., Simplicial and continuation methods for approximating fixed points and solutions to systems of equations, *SIAM Review*, **22**(1980), 28-85.

[3] de Branges, L., A proof of the Bieberbach conjecture, *Acta Math.*, **154**(1985), 137-152.

[4] Chow, S. N., Mallet-Paret, J. & Yorke, J. A., A homotopy method for locating all zeroes of a system of polynomials, in *Functional differential equations and approximation of fixed points*, H.-O., Peitgen & H.-O., Walther(eds), Springer Lecture Notes in Math., No. 730, 1978, 77-88.

[5] Christenson, C. & Voxman, W., *Aspects of topology*, Marcel Dekker, New York, 1977.

[6] Duren, P., Coefficients of univalent functions, *Bulletin Amer. Math. Soc.*, **83**(1977), 891-911.

[7] Eaves, B. C., A short course in solving equations with PL homotopies, *SIAM-AMS Proc.*, **9**(1976), 73-143.

[8] Eaves, B. C. & Scarf, H., The solution of systems of piecewise linear equations, *Math. of Op. Res.*, **1**(1976), 1-27.

[9] Gao, T. & Wang, Z., On the number of zeroes of exponential systems, *J. of Comp. Math.*, **9**(1991), 256-261.

[10] Gao, T. & Wang, Z., Several results on the number of zeroes of nonlinear mappings on bounded regions, *Chinese Quarterly J. of Math.*, **7**(1992), 44-47.

[11] Garcia, C. B. & Li, T. Y., On the number of solutions to polynomial systems of equations, *SIAM J. of Numer. Anal.*, **17**(1980), 540-546.

[12] Garcia, C. B. & Zangwill, W. I., Determining all solutions to certain systems of nonlinear equations, *Math. of Op. Res.*, **4** (1979a), 1-14.

[13] Garcia, C. B. & Zangwill, W. I., Finding all solutions to polynomial systems and other systems of equations, *Math. Prog.*, **16**(1979b), 159-176.

[14] Garcia, C. B. & Zangwill, W. I., An approach to homotopy and degree theory, *Math. of Op. Res.*, **4**(1979c) 390-405.

[15] Hayman, W., *Multivalent functions*, Cambridge Univ. Press, Combridge, England, 1958.

[16] Hille, E., *Analytic function theory II*, Ginn, Boston, 1962.

[17] Hirsch, M. W. & Smale, S., On algorithms for solving $f(x) = 0$, *Commu. Pure and Appl. Math.*, **32**(1979), 281-312.

[18] Jacobson, N., *Basic algebra I*, Freeman, San Francisco, 1974.

[19] Jenkins, J., *Univalent functions and conformal mapping*, Springer, New York, 1965.

[20] Kuhn, H., A new proof of the fundamental theorem of algebra, in *Mathematical programming study 1*, edited by M. L. Balinski, North-Holland, Amsterdam, 1974.

[21] Kuhn, H., Finding roots of polynomials by pivoting, in *Fixed points : algorithms and applications*, edited by S. Karamardian, Academic Press, New York, 1977.

[22] Kuhn, H., Wang Z. & Xu S., On the cost of computing roots of polynomials, *Math. Prog.*, **28**(1984), 156-163.

[23] Li, T. Y., On locating all zeroes of an analytic function within a bounded domain, *SIAM J. Numer. Anal.*, **20**(1983), 865-871.

[24] Li, T. Y., Sauer, T. & Yorke, J. A., Numerical solution of a class of deficient polynomial systems, *SIAM J. Numer. Anal.*, **24**(1987a), 435-451.

[25] Li, T. Y., Sauer, T. & Yorke, J. A., The random product homotopy and deficient polynomial systems, *numer. Math.*, **51**(1987b), 481-500.

[26] Li, T. Y. & Wang, X., Solving deficient polynomial systems with homotopies which keep the subschemes at infinity invariant, *Math. Comp.*, **56**(1991), 693-710.

[27] Ortega, J. M. & Rheinboldt, W. C., *Iterative solutions of nonlinear equations in several variables*, Academic Press, 1970.

[28] Rabinowitz, P. H., A note on topological degree theory for holomorphic maps, *Israel J. of Mathematics*, **16**(1973), 46-52.

[29] Renegar, J., On the cost of approximating all roots of a complex polynomial, *Math. Prog.*, **32**(1985a),319-336.

[30] Renegar, J., On the complexity of a piecewise linear algorithm for approximating roots of complex polynomials, *Math. Prog.*, **32**(1985b), 301-318.

[31] Renegar, J., A polynomial-time algorithm, based on Newton's method, for linear programming, *Math. Prog.*, **40**(1988a), 59-93.

[32] Renegar, J., Rudiments of average case complexity theory for piecewise linear path following algorithms, *Math. Prog.* **40**(1988b), 113-163.

[33] Royden, H., *Real analysis*, Macmillan, New York, 1968.

[34] Rudin, W., *Real and complex analysis*, MacGraw-Hill, New York, 1974.

[35] Scarf, H. E., The approximation of fixed points of a continuous mapping, *SIAM J. of Appl. Math.*, **15**(1967), 1328-1343.

[36] Shub, M. & Smale, S., Computational complexity : On the geometry of polynomials and a theory of cost : part II, *SIAM J. Computing*, **15**(1984), 145-161.

[37] Shub, M. & Smale, S., Computational complexity : On the geometry of polynomials and a theory of cost : part I, *Ann. Scient. Ec. Norm. Sup. 4 serie.* **18**(1985), 107-142.

[38] Smale, S., The fundamental theorem of algebra and complexity theory, *Bulletin AMS*, **4**(1981), 1-36.

[39] Smale, S., On the average speed of the simplex method of linear programming, *Math. Prog.*, **27**(1983), 241-262.

[40] Sternberg, S., Lectures on differential geometry, Prentice -Hall, 1964.

[41] Todd, M. J., *The Computation of fixed points and applications*, Springer-Verlag, Berlin, 1976.

[42] Van der Waerden, B. L., *Modern algebra, I, II*, Ungar publishing, New York, 1931.

[43] Walker, R. J., *Algebraic curves*, Princeton Univ. Press, Princeton, 1950.

[44] Wang, Z., On the efficiency of Kuhn's root-finding algorithm, *Kexue Tongbao*, **27**(1982), 1023.

[45] Wang, Z., *Fundamentals of Simplicial Fixed Point Algorithms*, Press of Zhongshan University, Guangzhou, 1986. (in Chinese)

[46] Wang, Z., A constructive approach to zero distribution of a class of

continuous functions, *Science in China* (Series A), **32**(1989), 1281-1288.

[47] Wang, Z., On the geometry of paths generated by PL homotopy Methods, *Ann. Op. Res.*, **24**(1990), 261-271.

[48] Wang, Z., Triangulate flat cones on simplices, *Acta Math. Sinica,* New Series, **7**(1991), 1-3.

[49] Wang, Z. & Xu, S., Approximate zeroes and computational complexity theory, *Scientia Sinica* (Series A), **17**(1984), 566-575.

[50] Xu, S., The GEK_k iterative algorithm and the geometry of polynomials, *Numer. Math.:A J. of Chinese Univ.*, **6**(1984), 345-354. (in Chinese, with abstract in English)

[51] Xu, S. & Wang, Z., Homotopy methods for systems of algebraic equations work with probability one, *J. China Univ. of Science and Technology,* **14** (1984),1, 15-22. (in Chinese, with abstract in English)

[52] Wang, Z. & Xu, S., A geometric proof on the monotonicity of Kuhn's algorithm approximating a simple root of polynomials, *J. China Univ. of Science and Technology,* **15** (1985),1, 19-29.

[53] Xu S. & Wang Z., The monotone problem in finding roots of polynomials by Kuhn's algorithm, *J. of Comput. Math.,* **1** (1983), 203-210.

Index

Acknowledgments

The authors are deeply indebted to Prof. W.-C. Hsiang, Prof. H. W. Kuhn, and Prof. Wen-tsun Wu for their encouragements and helps. In some sense, the monograph is a results of the authors' research under the guidance of Prof. H. W. Kuhn. They are grateful to Professors S. N. Chow, B. C. Eaves, C. B. Garcia, M. W. Hirsch, H. W. Kuhn, T. Y. Li, J. Mallet-Paret, H. Scarf, M. Shub, S. Smale, M. J. Todd, J. A. Yorke and W. Zangwill for kindly allowing them to reproduce some of their works. They are also grateful to China University of Science and Technology for offering the original Chinese publication of this monograph the first award of excellent graduate textbooks.

The works underlying the writing of this monograph are partially supported by the Foundation of Zhongshan University Advanced Research Centre, the Foundation of the National Committee of Education of China, and the National Natural Science Foundation of China.